Marine Engineering Systems

Marine Engineering Systems

An Introduction for Merchant Navy Officers

A. J. WEDDLE,
C.ENG., F.I.MAR.E., M.INST.M.C.

Extra First Class Engineer

Senior Lecturer
in Marine Systems
Plymouth Polytechnic

HEINEMANN : LONDON

William Heinemann Ltd
15 Queen St, Mayfair, London W1X 8BE

LONDON MELBOURNE TORONTO
JOHANNESBURG AUCKLAND

First published 1976
434 92233 1

Filmset in 'Monophoto' Times 11 on 13 pt by
Richard Clay (The Chaucer Press), Ltd, Bungay, Suffolk
and printed in Great Britain by
Fletcher & Son Ltd, Norwich

Foreword

by Capt. G. R. Hughes
Assistant Director, Plymouth Polytechnic

Shipping, perhaps more than any other industry, is experiencing the effects of the so-called technological revolution. Twenty years ago few people could have envisaged the half-million-ton V.L.C.C., the fast container ship, the liquefied natural gas or the chemical carrier, and few would have foreseen that the size of ships' crews would diminish rather than grow as the size and complexity of ships increased.

The need for efficient ship operation has meant an ever increasing demand for systematizing machinery and equipment functions, and the growth of systems that permit the removal of scarce, qualified manpower from operations that can be controlled just as well, or perhaps better, by machines. Department of Trade examinations for certificates of competency for deck and engineer officers now call for a knowledge of the general principles of systems engineering and of the shipboard applications of systems, and this book has been written primarily with the needs of students for certificates of competency in mind. However, most serving ship's officers, including those with their professional examinations behind them, also need some knowledge of systems if they are to understand their working environment.

The author of this book is an experienced and highly qualified marine engineer, who has spent many years teaching the subject to marine engineer and deck officers at all levels, studying for certificates of competency, for national certificates and diplomas, and for degrees in nautical science. He is therefore well aware of the need for an introductory book which is sufficiently comprehensive to cover ship systems in general but which has a nonmathematical treatment. The aim of this book is to present the subject in a readable form, supported by a large number of diagrams to illustrate the essential principles.

Preface

This book was written primarily for students preparing for the Department of Trade Certificate of Competency as Foreign Going Master, and for the Higher National Diploma in Nautical Science.

Since deck officers are concerned primarily with the response of ship systems, and modern ships are characterized by the interaction of power plant and measurement and control systems, the aim of this book is to encourage a broad view of the systems concerned which is not obscured by inappropriate attention to detail.

This approach is also particularly relevant to cadet and junior engineer officers, and marine electronics officers, who require a broad introductory approach to marine systems as a prerequisite to further studies in depth, which are a requirement of their particular development of competence.

My thanks are due to the Editor of *Safety at Sea International* for his permission to reproduce certain material which has appeared in his magazine, and to Mrs Betty Reynolds for typing the manuscript.

AJW

Contents

Chapter 1

Introduction: The Systems Approach

Modern ships are characterized by the presence of many protective devices and control systems which are extra to the basic power plant. Hence the reliability of the power-producing system, and of any other automated system, depends not only on the basic power-producing system but also on the many interacting elements of the complete system. This makes it necessary for those using the systems on board to be aware of the general principles by which the systems operate, in order that they may develop confidence in their use and quickly recognize changes in system response. Such early recognition can then lead to corrective action and so maintain and improve the safety of the vessel and those on board.

A system is an array of interacting component subsystems which co-operate to accomplish an objective. Figure 1.1 illustrates

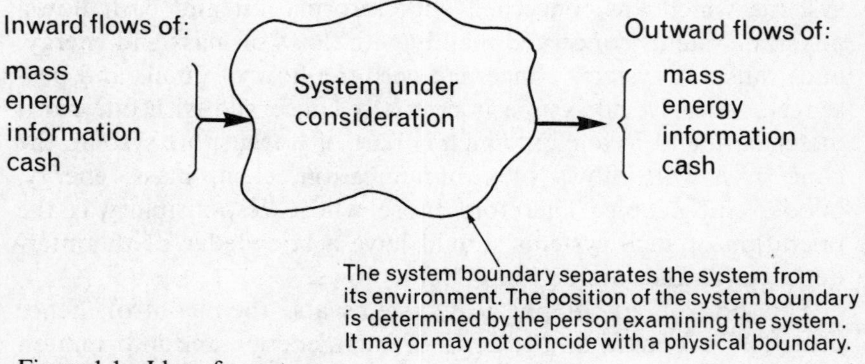

Figure 1.1 Idea of a system.

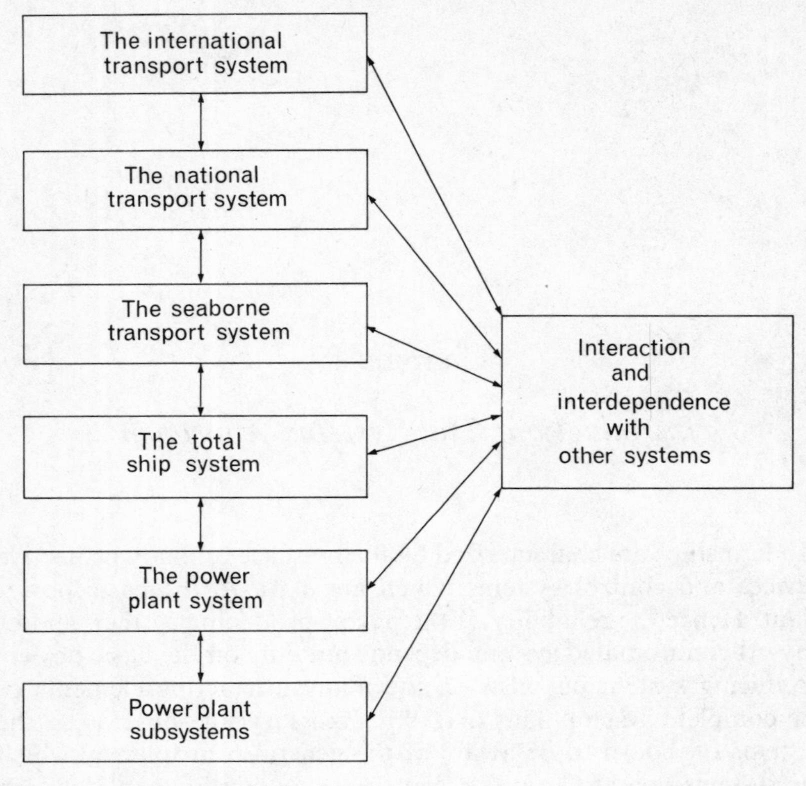

Figure 1.2 System relationships.

the broad idea of a system and Figure 1.2 the interrelationships which exist in a particular system hierarchy. Sufficient generality is maintained by this approach to cover, for example, organizational systems which are concerned with information and cash flows, physical systems concerned mainly with flows of mass and energy, and transport systems concerned with the flow of goods and passengers. Clearly, no system is normally concerned with one parameter only and in a ship, which is part of a transport system, the concern is with flows of communication, cash, mass, energy, goods, and people. Therefore those whose responsibility is the operation of such systems should have a knowledge of the interdependence and interaction of these flows.

This book is about physical systems and their control; hence attention is focused on the flows of mass, energy, and information in marine systems.

Chapter 2

Energy Conversion Systems

HEAT ENGINES

The movement of a vehicle submerged in, floating on, or skimming over the top of water requires that a force be applied to that vehicle. This force has its source in some kind of power plant, which may be a steam or gas turbine or a steam or gas reciprocating engine. Whichever plant is used, certain ideas and terminology are employed to describe its performance. The plant which provides the power for propulsion, electrical power, and all ship services is called a heat engine.

The function of a heat engine is to convert the energy, supplied to it during a transfer of heat energy, into some form of mechanical energy which can be used to drive, say, a propeller or an alternator. The principle of such an engine is illustrated in Figure 2.1. Using the diagram the flows of energy can be balanced. It can be seen that a heat engine degrades energy, from the highly available and useful energy in the fuel to the almost totally unavailable energy in the volume of sea and air through which the ship has just passed, which are at a slightly higher temperature than they were before.

The efficiency of the energy conversion system is defined thus:

$$\text{Power plant efficiency} = \frac{\text{Flow of mechanical energy from plant}}{\text{Energy flow supplied to the plant}}$$

In broad terms:

steam turbine plants attain about 35% efficiency
gas turbine plants attain about 30% efficiency

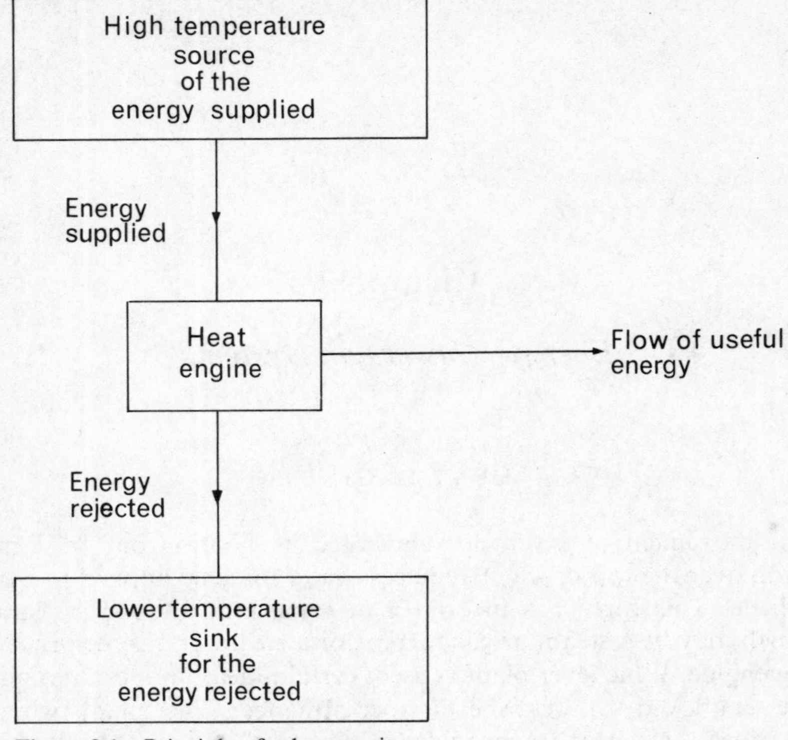

Figure 2.1 Principle of a heat engine.

steam reciprocating plants attain about 15% efficiency
gas reciprocating (i.e. diesel-engined) plants attain about 40% efficiency

The reasons for these apparently low conversion efficiencies are quite fundamental and no remarkable increases can be expected in the foreseeable future. Hence it can be seen that the world's heat engines are using up its 'capital' of easily available energy at an ever increasing rate, with a low conversion efficiency, resulting in an accelerating shortage of easily available energy.

THE FEED WATER SYSTEM AND ITS COMPONENTS

The feed water system allows for the continuous recycling of water and steam in a circuit and so enables water and steam to be used as carriers to transport energy from one part of a steam plant to another. In such a steam plant, energy is supplied to water and steam in a boiler from the combustion of fuel. This steam, with a high energy content, passes to a steam turbine which converts some of this energy into useful work. The exhaust steam then passes to the condenser where energy is rejected to the sink, which is the ocean. The energy and mass transfers in a typical feed system are shown in Figure 2.2.

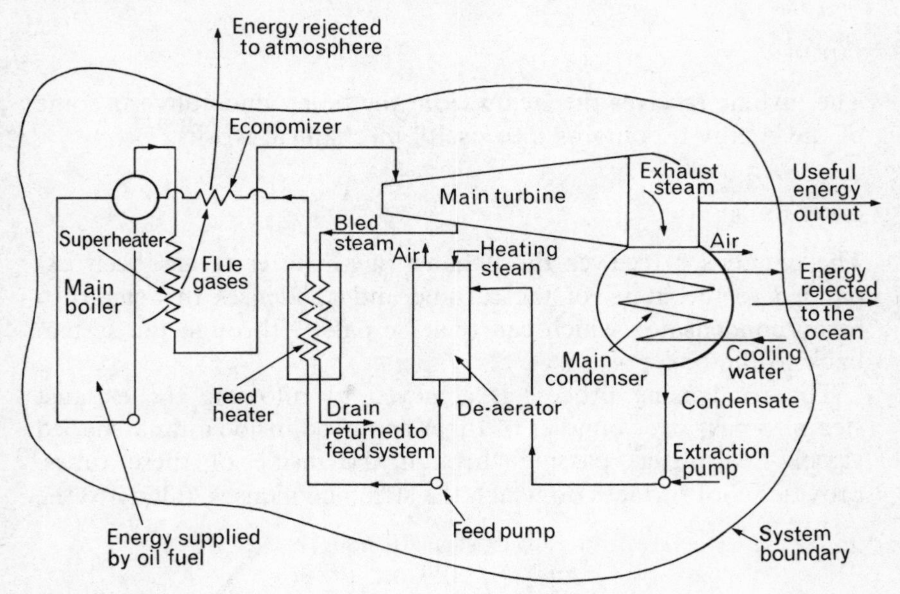

Figure 2.2 Feed water system.

The essential *functions* of the feed system are:

1. To transfer condensed steam from the condenser to the boiler. Clearly, this feed water has to come from somewhere, and if the exhaust steam were passed to the atmosphere the ship would have to carry a colossal amount of boiler feed water on board.
2. To enable the efficiency of the plant to be increased by using feed water heaters, which use steam bled from the turbines or exhaust steam from auxiliaries, to increase the temperature of the feed water before it enters the boiler.
3. To enable the condition of the feed water to be chemically analysed and controlled, and so to protect the boiler and the feed system from the effects of scale, oil, and air.

The essential *components* of the feed water system are listed below. In subsequent sections of this chapter some of these components will be examined more closely.

BOILER

The boiler transfers the energy released by the combustion of fuel to the water and steam contained within the boiler.

TURBINE

The turbine receives the steam from the boiler and converts some of the energy it contains into useful mechanical work.

CONDENSER

The condenser receives the exhaust steam after it has been expanded to the limits of the turbine and condenses this steam to form 'condensate', which can then be passed through the system back to the boiler.

The condensing process is achieved by allowing the exhaust steam to pass over bundles of tubes arranged inside a tank-shaped vessel. Sea water, passing through the inside of these tubes, provides cool surfaces on which the steam condenses, falling to the

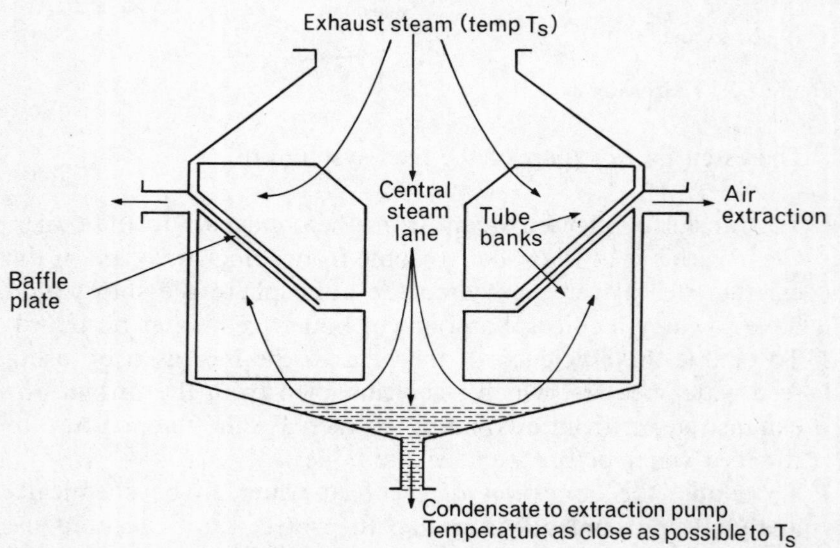

Figure 2.3 Regenerative condenser.

bottom of the condenser like rain. In a regenerative condenser the falling rain of condensate passes through exhaust steam, so that it leaves the condenser at a temperature closely approaching that of the exhaust steam. This ensures that:

1. The condensate is not undercooled, which would lead to a waste of energy in the plant.
2. The condensate contains a minimum of dissolved oxygen, which it absorbs from any air which has leaked into the plant.

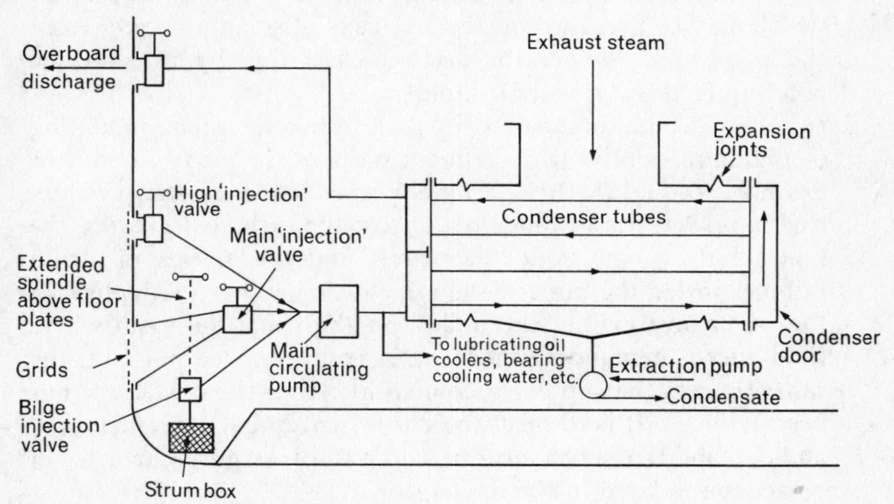

Figure 2.4 Condenser connections.

Figures 2.3 and 2.4 illustrate the condenser and its circulating connections.

The *bilge injection valve* is an important fitting, which has on occasions saved a ship by preventing the flooding of the engine room. By closing other suction valves, the large circulating pump can be used to pump water entering the engine room through the condenser and overboard. By this means it may be possible to maintain power supplies until the ingress of water causing the flooding can be controlled, or possibly the ship can reach harbour or be beached.

EXTRACTION PUMP

As steam condenses, its volume reduces by a factor of about 28 000. This great reduction in volume causes a vacuum to form in the condenser, which provides an increase in turbine power of about 20 per cent. It is important to maintain this vacuum, and the extraction pump helps by pumping out condensate from the condenser and delivering it to the de-aerator.

DE-AERATOR

The de-aerator fulfils two important functions:

1. It removes air from the condensate and any make-up feed water which may be necessary to replace leaks of steam or water from the plant. The de-aerator accomplishes this by causing the entering feed water to boil rapidly.
2. It fulfils the requirement of a surge tank by accommodating changing rates of flow in different parts of the feed system. For example, should the bridge watchkeeper request a sharp reduction in power, the steam demand from the turbine decreases, the fuel supply to the boiler decreases, and the masses of steam bubbles inside the boiler collapse, causing a sharp reduction in the water level within the boiler. So the situation arises where the boiler is demanding more water from the feed pump at the same time as the supply of condensate from the condenser has been reduced. It is to meet these transient imbalances in supply and demand that a reservoir of feed water is required, which can be accommodated in the de-aerator.

FEED WATER PUMP

The feed water pump increases the pressure of the feed water so that it may move on through the feed system and enter the boiler.

FEED WATER HEATERS

The feed water heaters utilize partially used steam, bled from the main turbines, or exhaust steam, to raise the temperature of the feed water before it enters the boiler; so they help to increase the efficiency of the power plant.

MARINE STEAM BOILERS

The feed water entering the boiler from the feed system has a relatively low energy content compared with the steam leaving the boiler. This increase in energy is supplied by burning fuel within the boiler, the effectiveness of this heat energy transfer being described by the boiler efficiency.

$$\text{Boiler efficiency} = \frac{\text{Energy transferred to the feed water in producing steam}}{\text{Energy released by the combustion of fuel within the same time period}}$$

The fuel may be in solid form, or liquid, or gaseous. The majority of power station boilers use solid fuel in the form of coal, boilers in steelworks often use blast furnace gas, and the vast majority of marine boilers use oil fuel. In a nuclear fuelled ship or power station a nuclear reactor is used to transfer energy to the feed water to produce steam, the rest of the feed system remaining fundamentally the same. A small conventional marine boiler is supplied as a 'get you home' facility.

The boiler's function is to generate steam from water. Therefore it must consist of:

1. A furnace or combustion chamber in which fuel may be burned efficiently, releasing energy.
2. Heat transfer surfaces which allow the energy, so released, to reach the fluid contained within the boiler and so to form steam.
3. A chamber to allow steam and water to separate.
4. Passages or ducts which allow for the supply of air to burn the fuel and for the escape of flue gases to the atmosphere.
5. Various fittings and control devices which ensure that the supplies of fuel, air, and feed water are balanced with the demand for steam.
6. Various fittings concerned with the safe operation of the boiler, e.g. safety valves and fuel shut-off valves.

Boilers may be divided into two broad classes: water-tube boilers and smoke-tube boilers.

Water-tube Boilers

These are so called because the hot gases resulting from the combustion of fuel surround the heat transfer surfaces, which consist largely of steel tubes about 65 mm in diameter. These tubes contain boiling water and steam bubbles under pressure. Since the tubes are small in diameter, with a wall thickness of about 5 mm, they can withstand fairly high pressures under operating temperatures, and the wall thickness allows for effective heat transfer. A marine water-tube boiler is shown in Figure 2.5.

Standing inside the furnace is comparable to standing in a tall room which has its floor, ceiling, and all its walls covered with tubes. In modern boilers the spaces between the tubes are filled by strips of steel, welded into position. By this means the complete furnace is a steel box, which has its gas-tight surfaces cooled by the steam and water mixture circulating through the tubes. The tubes

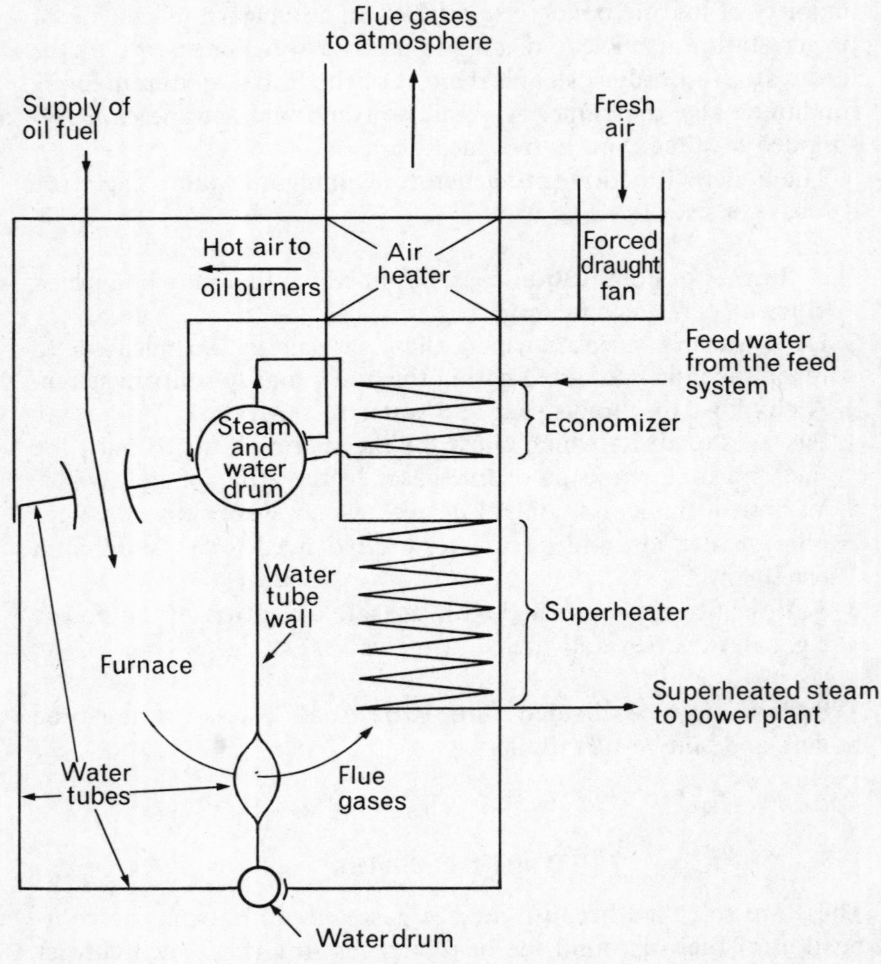

Figure 2.5 Marine water-tube boiler.

are bent in the roof of the furnace, to allow the insertion of the oil-burning assembly. Near the foot of the water-tube wall each successive tube is bent in the opposite direction, which allows the hot gases resulting from the combustion of fuel to leave the furnace. The flow of hot gas is now directed over other heat transfer surfaces, which extract the maximum possible heat from the flue gases before they pass out of the furnace to the atmosphere.

The water tubes are fitted into cylindrical pressure vessels called 'drums', the steam and water drum fulfilling the function of a separation chamber.

The circulation of the steam and water mixture within the furnace tubes is maintained by the presence of large-diameter tubes which pass from the top to the bottom drum, outside the furnace. Since these tubes receive no heat energy no steam is generated within them, as is the case in those heated tubes inside the boiler which contain large numbers of steam bubbles.

Clearly, the contents of the unheated 'downcomers' exert a greater pressure at the water drum than do the contents of the heated 'risers', and it is this difference in pressure which causes the circulation of fluid, which keeps the tubes at a safe working temperature. If the water level within the steam and water drum drops below a certain limit this circulation is seriously affected, which causes the tube material to approach the gas temperature. In this case the tube material cannot withstand its internal pressure at such elevated temperatures and tube failure results.

Properties of Steam

The diagram of the water-tube boiler may be used to introduce some of the definitions used by engineers to describe certain properties of steam.

Water entering the economizer from the feed system increases in temperature as it gains energy from the flue gases passing over the economizer tubes. This gain in energy is called gain in liquid enthalpy. Since the water is under pressure in the economizer, its boiling temperature is much greater than the boiling temperature of water at atmospheric pressure, which is 100 °C.

The water entering the steam and water drum must first have its temperature raised to that unique boiling, or 'saturation', temperature associated with the pressure in the boiler drum. Then further energy must be supplied to change the boiling water, at saturation temperature and pressure, to dry saturated steam, at the same saturation temperature and pressure. This further amount of energy is called the enthalpy of evaporation.

Since the steam so formed is dry, and is at its saturation temperature, it is called dry saturated steam. Steam which is at the saturation temperature, but which still carries in suspension some particles of unevaporated boiling water, is called wet steam.

If dry saturated steam is removed from the water from which it has been formed and further energy is supplied to it, its temperature rises above the saturation temperature corresponding to its pressure. Then the steam is said to be superheated. The energy supplied is called the enthalpy of superheat, and that part of the boiler allocated to this process is called the superheater. Superheating is used because it reduces the size and weight of the plant necessary to produce a required power, and in order to increase the thermal efficiency of the plant.

Figure 2.6 further illustrates these ideas.

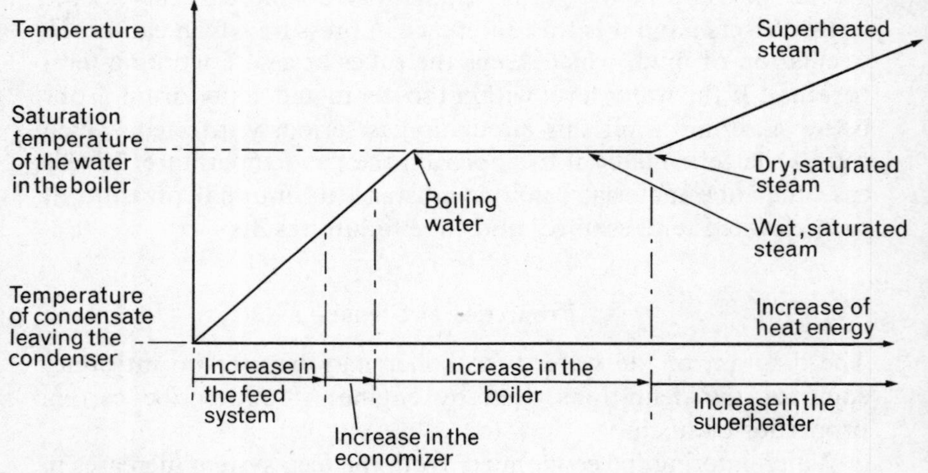

Figure 2.6 Properties of steam.

Smoke-tube Boilers

In these boilers the hot gases resulting from combustion are contained within the heat transfer surfaces.

Figure 2.7 shows a vertical cross-section through a cylindrical 'Scotch' marine boiler. Since the heat transfer surfaces contain the hot gases, in turn the steam and water must be contained in a

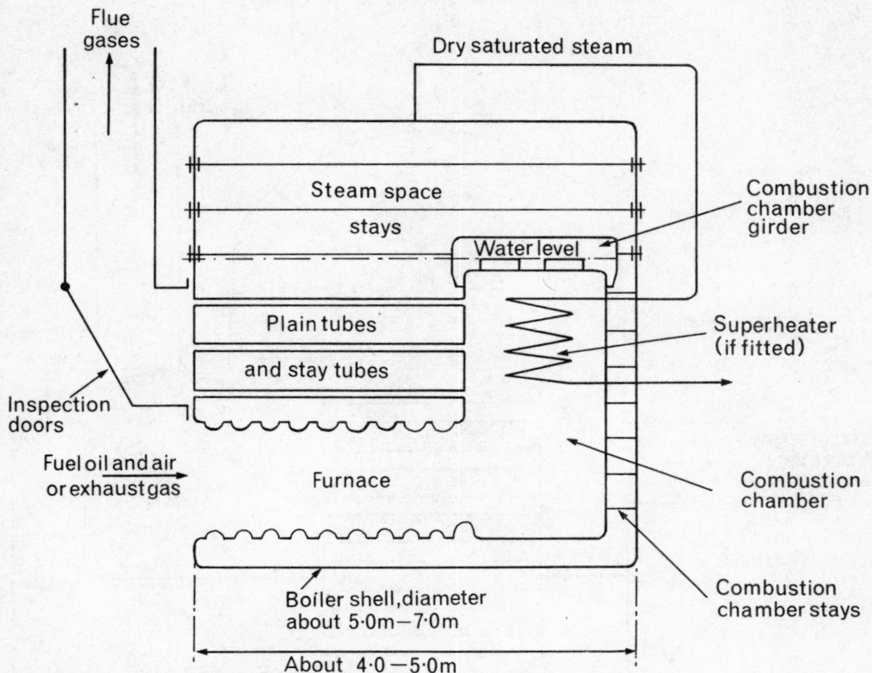

Figure 2.7 'Scotch' marine boiler.

pressure vessel, which surrounds both the steam and water mixture and the heat transfer surfaces. Therefore, if the size of these surfaces is increased, the diameter of the pressure vessel also increases, which limits the working pressure of the boiler to about fifteen times atmospheric pressure. These limitations restrict the power available from a Scotch boiler to about 3 000 kW, which is about one-tenth of the power available from the water-tube boiler described above. Because of this power limitation Scotch or similar boilers are rarely used for main propulsion purposes. They find general use as auxiliary boilers in motor ships, making use of the large amount of energy available in the exhaust gas of diesel engines.

In some auxiliary boilers the cylindrical pressure-containing vessel is arranged vertically, with the tubes and furnace arranged as indicated in Figure 2.8. These boilers have an even smaller output than the Scotch boiler and supply steam essentially for port services.

There is an accelerating trend for large diesel engines to be fitted with very sophisticated energy-recovery arrangements, due to the recent dramatic increase in oil prices. Many of these set-ups use

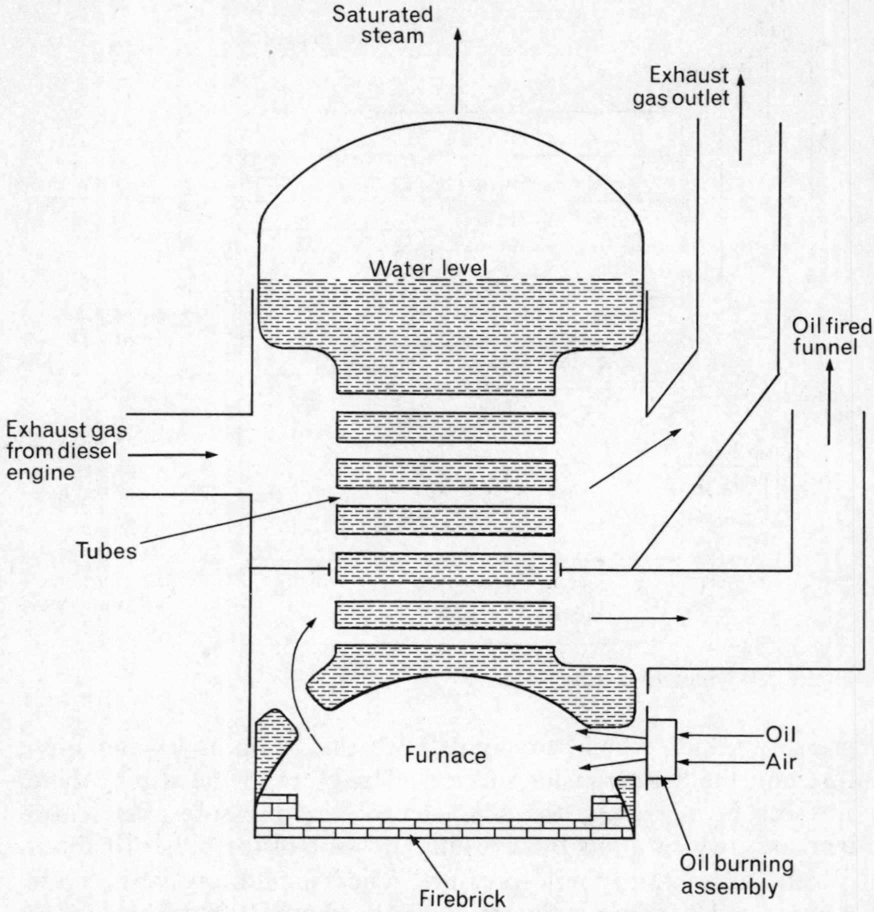

Figure 2.8 Composite auxiliary boiler arranged for exhaust gas and oil fuel firing.

dual pressure heat-recovery systems incorporating high-pressure heat exchangers in the flue gases, transferring energy to a water to steam separation vessel. A block diagram of such a system is shown in Figure 2.9; the subsystems are described later in this chapter.

Preparation of a Boiler for Steaming

When metals are heated they expand, and if one part of a metallic structure is heated at a faster rate than another part the differential expansion causes stresses to be set up in the material, which may reach unacceptable levels.

The diagram of the Scotch boiler (Figure 2.7) shows that an attempt at rapid steam raising would cause the top of the furnace to become hot very quickly, while the stagnant water lying underneath the furnace would keep the lower part of the furnace relatively cool. Such differential expansion would induce high stresses which might result in distortion of the furnace. To prevent this, steam must be raised very slowly in a smoke-tube boiler. Typically, nearly a day is required for a cold system.

Steam may be raised in a water-tube boiler much more quickly because of:

1. Its lower mechanical stiffness, which allows expansion to be accommodated without its unduly raising stress levels.
2. Its much greater degree of natural circulation.

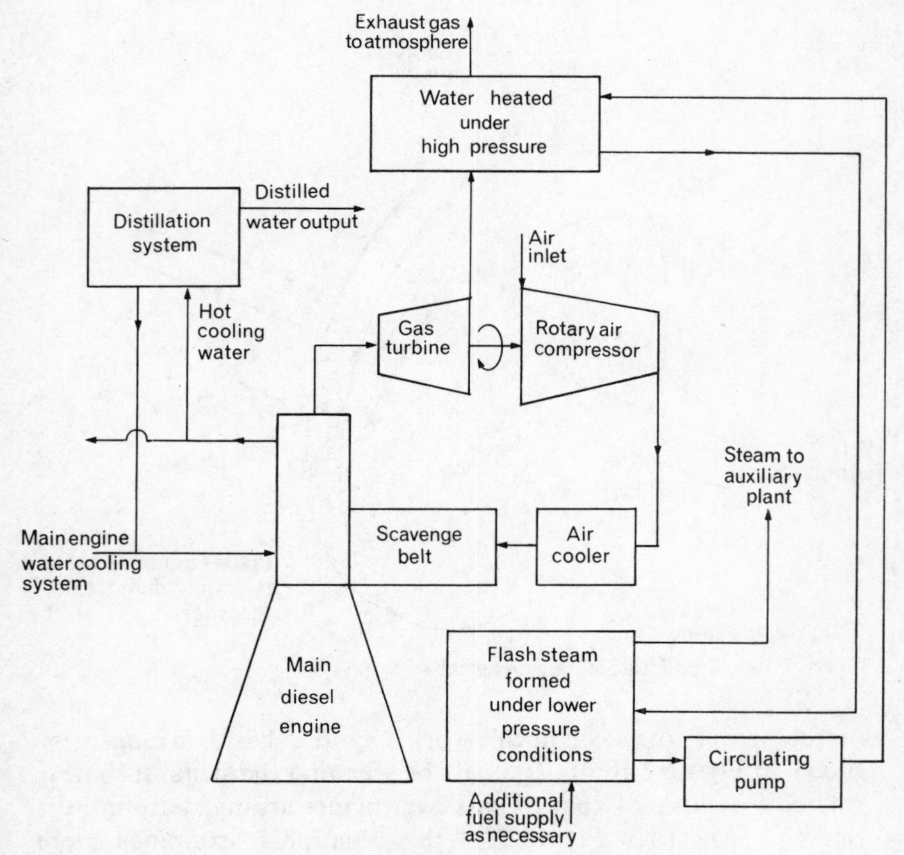

Figure 2.9 Energy recovery systems in a diesel installation.

A typical time to raise steam in a water-tube boiler may be about six hours, but each installation has its own requirements.

In every type of boiler, particular care is taken to protect superheaters when raising steam, by maintaining a cooling flow through them. Those reading this book may have heard the superheater vents in the funnel whistling away, venting cooling steam during the steam-raising procedure.

MARINE STEAM TURBINES

The steam entering the turbine from the boiler has a high energy content, compared to the steam exhausted from the turbine to the condenser. The turbine converts some of this energy to provide a

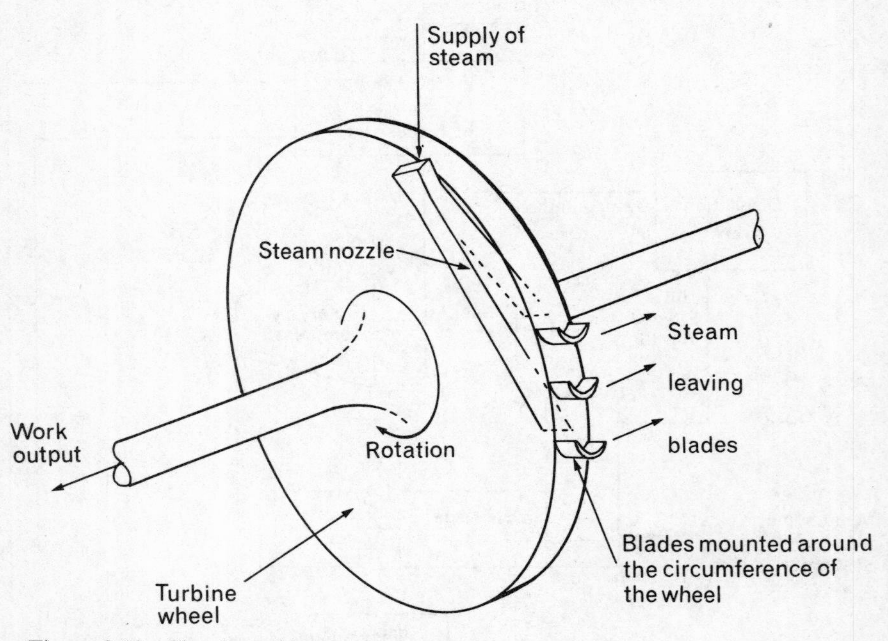

Figure 2.10 Simple turbine arrangement.

useful source of mechanical work, by the basic arrangement shown in Figure 2.10. It can easily be seen that the steam turbine is rotated by causing steam to pass over blades arranged around the rim of a wheel. In Figure 2.11 the situation is examined more closely.

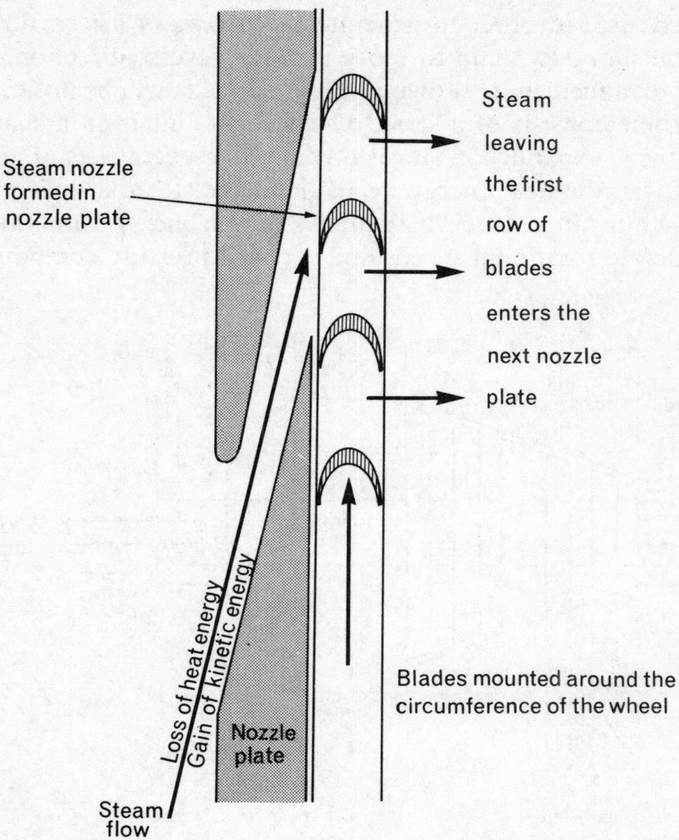

Figure 2.11 Arrangement of nozzles and blades in a turbine.

Applying a principle learned in mechanics:

Force =	Mass	× Acceleration
Force on blade =	Mass flow of steam	× Change in velocity of steam
Units: [N =	kg/s	× m/s]
or [N =	kg (Mass)	× m/s² (Acceleration)]

Hence, it can be seen that by allowing steam to blow across a curved blade, and so to change its direction and speed, a force is produced which will tend to turn the wheel and hence to drive a propeller or pump through an appropriate system. However, these diagrams clearly show that the turbine wheel will rotate in one direction only. This is acceptable for driving pumps and alternators, but for driving a ship's propeller an 'astern' turbine is required also, in order to generate the necessary astern thrust to slow the ship down and to move it in the reverse direction.

This arrangement is shown in Figure 2.12. It will be noticed that the turbine consists of a series of wheels mounted on a shaft, the steam being expanded in stages through the several sets of nozzles. In this way the heat energy available in the steam is split up into stages, an arrangement which allows the turbine to run efficiently at moderate rotational speeds and is called pressure compounding

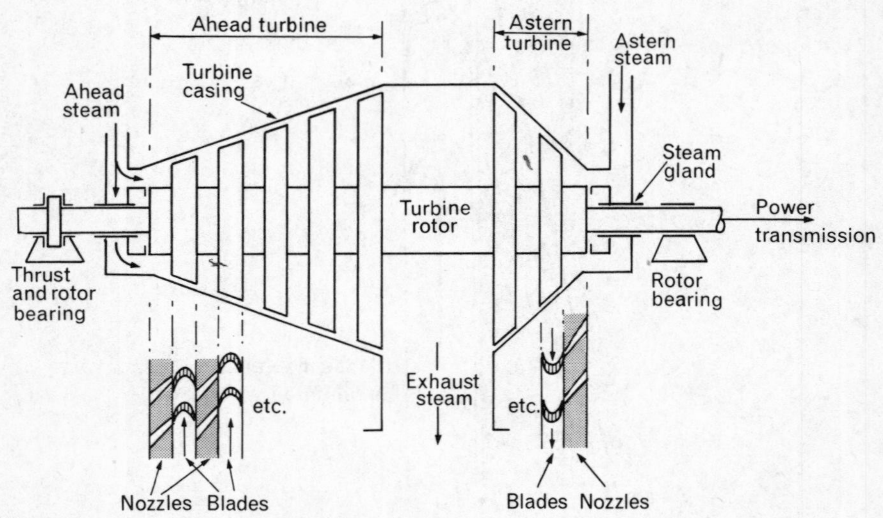

Figure 2.12 Arrangement of a steam turbine.

the turbine. In small turbines the steam may be expanded through a large pressure drop, resulting in very high-velocity steam speed, the subsequent use of which is split into stages. This arrangement is called velocity compounding.

By opening and closing control valves, known as manoeuvring valves, at the inlets to the ahead and astern turbines, steam at various flow rates can be admitted to obtain the required amount of power in the appropriate direction. When the turbine is running ahead, the astern turbine acts as a compressor, which absorbs some of the turbine power. This is the basic reason why the astern turbine is kept as small as is possible but consistent with ship safety; it develops about 50 per cent of the power of the ahead turbine. If the turbine is moving astern the ahead turbine also acts as a compressor, and if the turbine runs full astern for more than a limited period the power absorbed by the ahead turbine in compressing the steam causes overheating of the turbine. For this reason a time limit on astern running is usually imposed on a turbine ship, typically thirty minutes at full astern revolutions.

Operating Procedure for a Steam Turbine

PREPARATION FOR SEA

It will be appreciated from Figure 2.12 that the turbine rotor is a massive unit, supported on widely separated bearings which have very small internal clearances between the rotating turbine wheels and the stationary nozzle plates. Hence, preparation for sea is aimed at the gradual warming through of all the component parts of the turbine, avoiding thermal stresses and distortion, while maintaining the necessary clearances. These aims are attained by the following general procedure, which may differ, in detail, from the requirements laid down by a specific turbine manufacturer.

1. All steam pipe and turbine drains are opened.
2. The lubricating oil pumps are started, and the lubrication system is checked through.
3. The circulating water pump is started, after the position of the overboard discharge has been checked. (Barges alongside have been sunk in a very short time by receiving this discharge.)
4. The propeller is checked for obstructions, e.g. ropes and barges, and the complete turbine installation, gears, and propeller shaft are turned very slowly by the turning gear.
5. Steam is admitted to the turbine through a small warming-through connection, which bypasses the main valves.
6. A small vacuum is raised, gland sealing steam is admitted, and the warming and turning process continues for a period which depends largely on the period of time the turbines have been shut down and on the ambient temperatures. Typically, this period is about two hours. The turbine is turned continuously to allow even heating of the rotor and so avoid thermal stresses.
7. The turning gear may now be removed. Main steam is admitted in very small quantities, to spin very slowly the turbines a few revolutions ahead and astern, the vacuum being raised to about 400 mm of mercury. This process is continued for about two hours. Near the end of the period the turbines are 'blasted'

ahead and astern. The vacuum is raised. Safety and control devices are checked.

8. The plant is now ready for 'stand by'; all pumps are running at the speeds necessary for manoeuvring. The engine may now be manoeuvred, but sympathetic pilots try not to ask for full ahead or full astern as the first movement of the plant. During the manoeuvring period, the turbine rotors must not remain stationary for more than a few minutes. If this were allowed to happen, the rotor could distort and the next movement would result in internal mechanical damage.

AT FULL AWAY

The full ahead and full astern manoeuvring revolutions of the plant are typically about 80 per cent of the 'full away' revolutions, so that the plant at full away may be producing only about half its full power. Hence, when full away is rung on the telegraph the engineer officers gradually increase the output of the plant and begin 'tuning' it for its maximum efficiency. This involves establishing certain services such as turbo-alternators, which use steam bled from the main turbines, and generally optimizing the many parameters which, when combined, result in an efficiently operating plant. The process typically takes another one and a half hours before the full speed of a large vessel is reached.

ARRIVAL AT THE NEXT PORT

Since the turbine requires an acceleration period, it follows that it also requires a deceleration period. Hence the engineering department is informed of an impending arrival about two hours before the first expected movement. The engineer officers can then begin to prepare the plant, transferring certain bled steam services to the main boiler and gradually reducing the power output of the plant, over a period of about one and a half hours, to manoeuvring full ahead revolutions.

EMERGENCY FULL ASTERN

Clearly, in an emergency, the previously described desirable procedure must be ignored. In such a situation, at emergency full astern, electrical power supplies are secured, the main steam ahead valve is closed, and astern steam is admitted very gradually as the engine slows down. As the ahead revolutions drop, the amount of astern steam is increased so that, after about five minutes, the turbine is rotating astern. The time intervals involved depend on the size and type of ship, and the reader is referred to the information on the stopping characteristics of the ship, which may be available on board. Figure 2.13 illustrates the relationship between

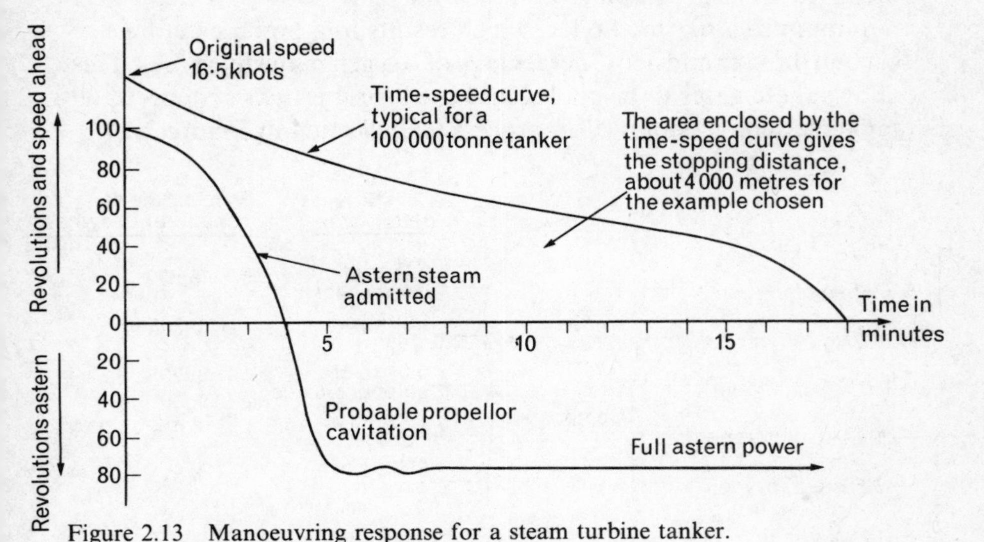

Figure 2.13 Manoeuvring response for a steam turbine tanker.

ship speed, power plant response, and head reach. Clearly, the use of the emergency or 'crash' astern manoeuvre increases the probability of damage to turbines, gears, and boilers.

THE COMBUSTION PROCESS AND FUEL SUPPLY SYSTEMS

Combustion Process

The great majority of marine power plants obtain their necessary supply of energy from the combustion of coal, fuel oil, or hydrocarbon gas. The chemical combination of the combustible elements in a fuel with oxygen in the air releases energy which can be utilized in the furnace of a boiler, the combustion chamber of a gas turbine, or the cylinder of an internal combustion engine.

The principal combustible constituents of solid, liquid, or gaseous fuels are hydrogen and carbon. The oxygen which is needed to burn these constituents is contained in the air of the atmosphere, which contains approximately 23 per cent oxygen and 77 per cent nitrogen by mass. The theoretical mass of air required to burn 1 kg of a certain fuel can be calculated, but in practice it is found necessary to supply air in excess of this requirement, in order to ensure complete combustion. The supply is kept to a minimum in a marine boiler, which results in a high percentage of carbon dioxide and a low percentage of oxygen in the flue gases. This allows these gases to be cooled and used for inert gas systems in oil tankers. Such a combustion process is illustrated in Figure 2.14.

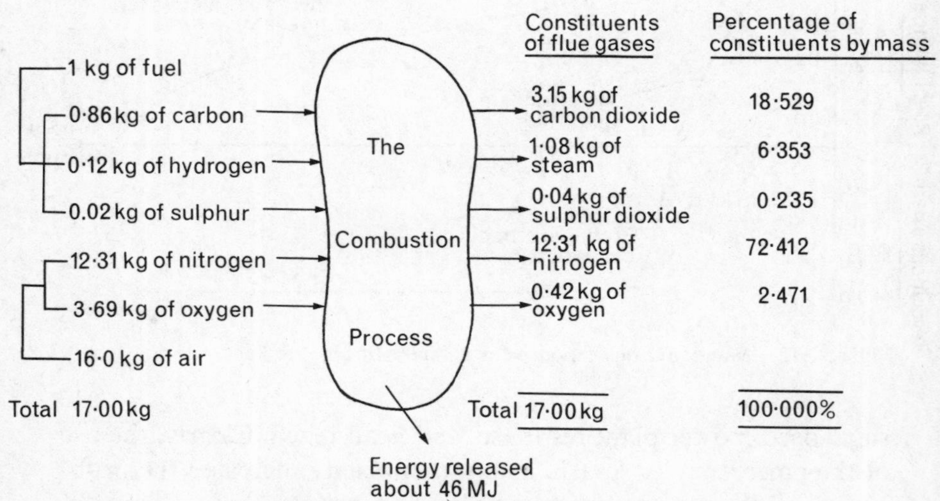

Figure 2.14 Combustion process of a marine boiler.

The diagram is not typical of diesel engines or gas turbines since both these use larger amounts of excess air, primarily to act as an internal coolant. As a result their flue gases may not be suitable for inerting systems, because of their higher content of oxygen.

Combustion Requirements

For *efficient combustion* two things must be ensured:

1. There must be sufficient oxygen present to completely burn the fuel.

2. The oxygen required must make contact with the combustible constituents while the fuel is in a combustible condition.

To encourage these factors:

1. The fuel is broken up into very small particles. This process exposes a very much greater surface area of fuel to oxygen, which encourages rapid combustion. (It may be relevant to remark here that any atmosphere containing combustible dust is potentially dangerous. In saw mills, flour mills, and coal mines special efforts are taken to minimize explosions due to this cause.)
2. The air and fuel are intimately mixed by ensuring that there is relative motion between them. This relative motion, called turbulence, removes the products of combustion from a droplet of burning fuel, so exposing the surface of the fuel to fresh air.

Oil fuel is split up into very small particles by forcing it through very small holes. If the fuel has the correct viscosity, it will issue from the holes in a finely divided spray of oil. The devices performing this function are called atomizers, burners, or fuel injectors. A typical oil fuel burner for a boiler is illustrated in Figure 2.15.

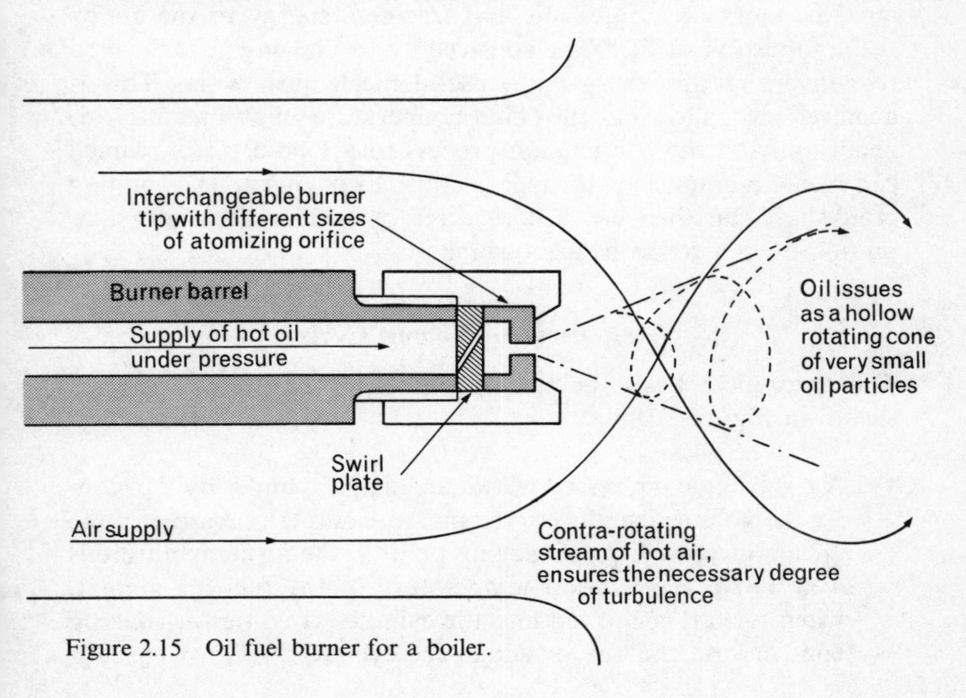

Figure 2.15 Oil fuel burner for a boiler.

Fuel Supply Systems

Typical fuel systems for steam and diesel installations are shown in Figures 2.16 and 2.17. In the diesel engine, fuel is burned within the engine cylinder, which requires fuel pumps and injectors with very fine mechanical clearances. Hence more effort is taken to remove undesirable abrasive constituents from the fuel oil. The boiler fuel system can tolerate much higher levels of these constituents. The boiler itself can accommodate higher percentages of other undesirables, e.g. vanadium and sulphur, and therefore boilers are able to burn very low-grade fuels successfully.

In case of fire, the valves controlling the flow of oil from the tanks shown in the diagrams can be closed remotely from the upper deck, and remote switches enable the pumps and fans to be stopped from a safe position outside the engine room.

MARINE DIESEL ENGINES

The diesel engine burns its fuel within the engine itself, transferring energy directly to the working fluid of the engine, which is air. The engine aspirates air, and transfers energy to the air by the combustion of fuel. The construction of the engine is devoted to converting this energy into useful mechanical work. This is achieved by allowing the high-temperature high-pressure air resulting from the combustion process to act on a piston, which can rotate a crankshaft through a crank mechanism. The rotating crankshaft can then be used to drive propellers, alternators, or pumps, as in a steam or gas turbine.

Basic Pressure–Volume Cycle

The theoretical basis for the operation of the diesel engine is shown in Figure 2.18.

1–2 Air is being compressed within an engine cylinder by a piston. As the volume is reduced, the pressure and temperature of the air are increased. Approaching point 2, the air temperature is well above the ignition temperature of the fuel oil so that, when fuel oil is injected into the cylinder, it ignites on making contact with the hot air within the cylinder.

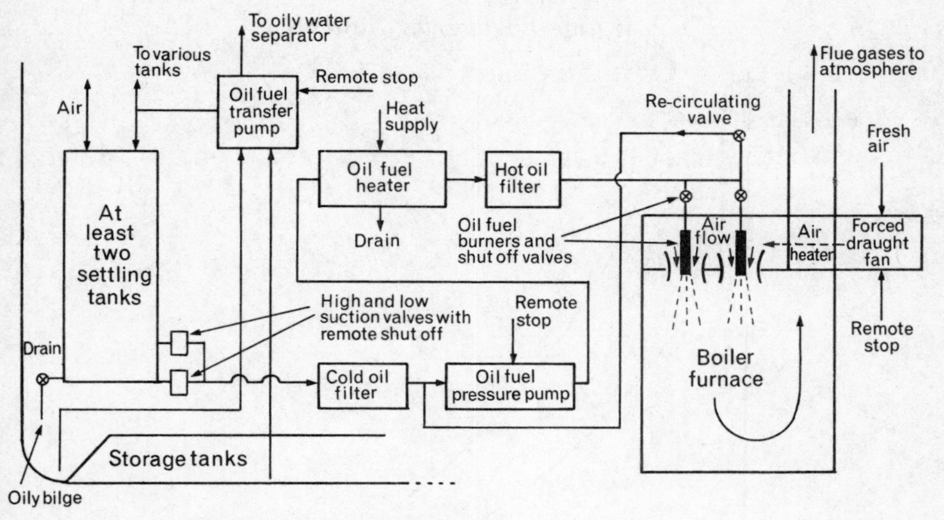

Figure 2.16 Steam plant fuel-supply system.

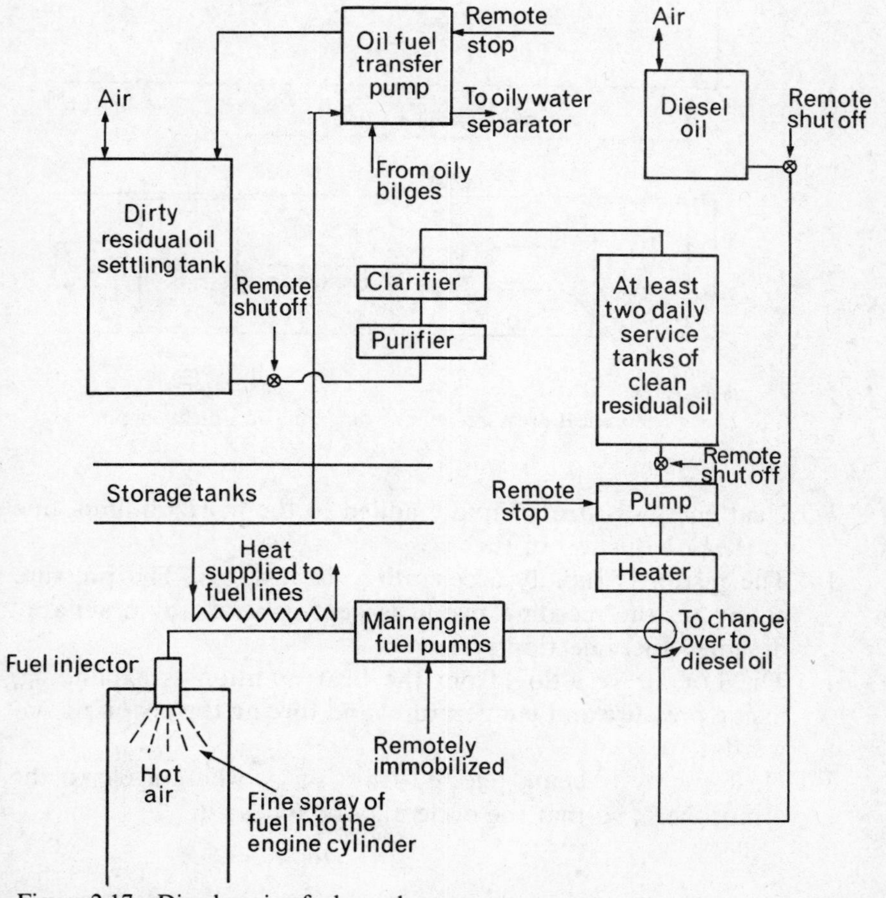

Figure 2.17 Diesel engine fuel-supply system.

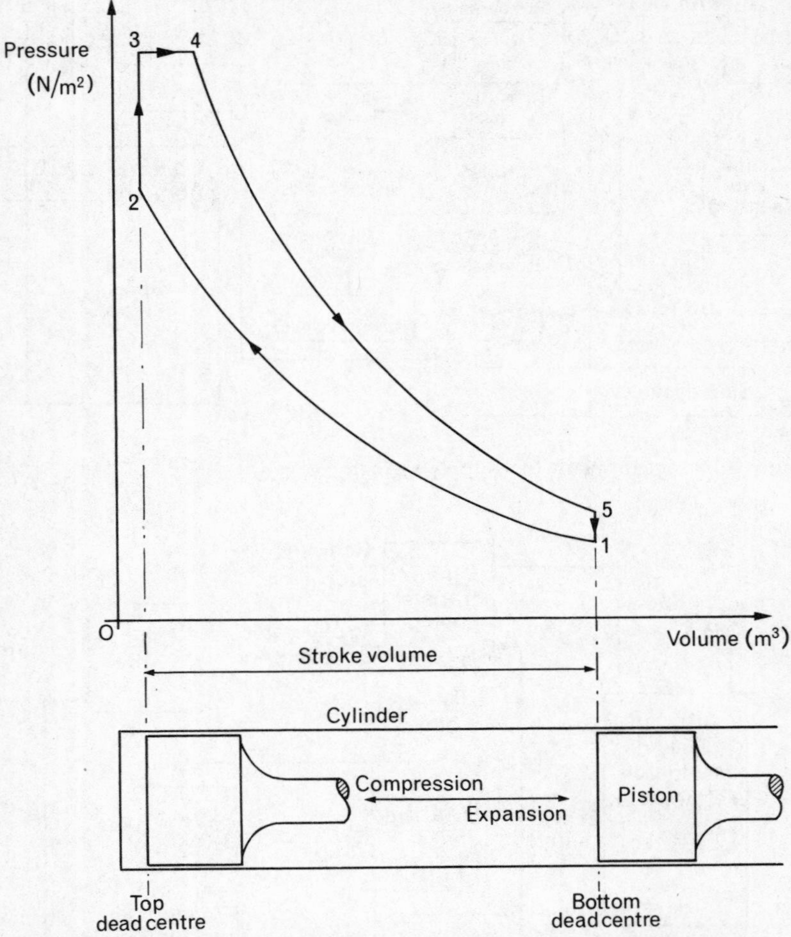

Figure 2.18 Theoretical pressure–volume diagram for a diesel engine.

2–3 Heat energy is being rapidly added to the working fluid, due to the combustion of fuel.

3–4 The piston is rapidly accelerating downwards. The pressure acting on the receding piston is kept constant by a suitably designed fuel injection system.

4–5 The hot air resulting from the heat addition is expanding, losing pressure and temperature, and forcing the piston downwards.

5–1 Heat energy is being rejected from the working cycle to the atmosphere, so that the cycle may be repeated.

Consider an element of the area of Figure 2.18. Height represents pressure in N/m^2 and length represents volume in m^3.

$$\text{Height} \times \text{Length} = \text{Area}$$

Thus the area represents units of:

$$[N/m^2 \times m^3 = Nm = J\ (\text{Joules})]$$

That is, the area of the diagram represents the amount of energy theoretically available from the engine during one cycle. Further reference to this point will be made in Chapter 6.

SUPERCHARGING

The area of the cycle diagram can be increased by increasing the pressure at point 1, while maintaining the same volume. This ensures that the mass of air following the cycle is increased, which allows more fuel to be efficiently burned in the cylinder. This results in a greater power output for a given size of engine.

The air may be supplied to the engine under pressure by one of two mechanisms:

1. An air compressor mechanically driven from the engine or other prime mover.
2. An air compressor, driven by a gas turbine, which is rotated by the exhaust gases leaving the engine. This arrangement, which is the more usual one, is called a turbocharger or turboblower.

Medium- and Low-Speed Diesel Engines

The two main classes of marine diesel engines operating on this cycle are medium-speed and low-speed engines.

MEDIUM-SPEED DIESEL ENGINES

Some of these work on the four-stroke process, which can be explained with reference to the diagram of the medium-speed engine (Figure 2.19) and to the four-stroke timing diagram (Figure 2.20).

1st stroke—the induction stroke. As the piston descends, the cylinder is filled with air through the inlet valve.

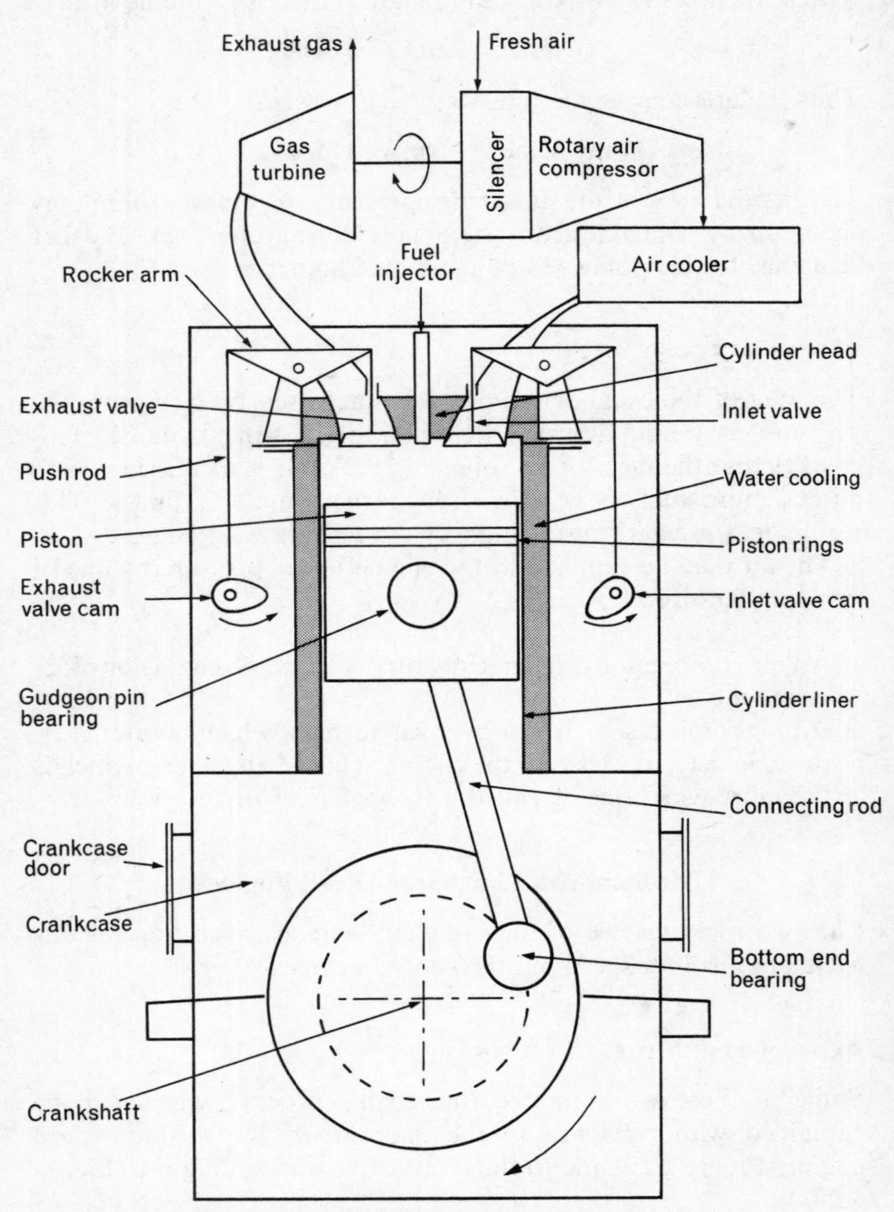

Figure 2.19 Four-stroke medium-speed diesel engine.

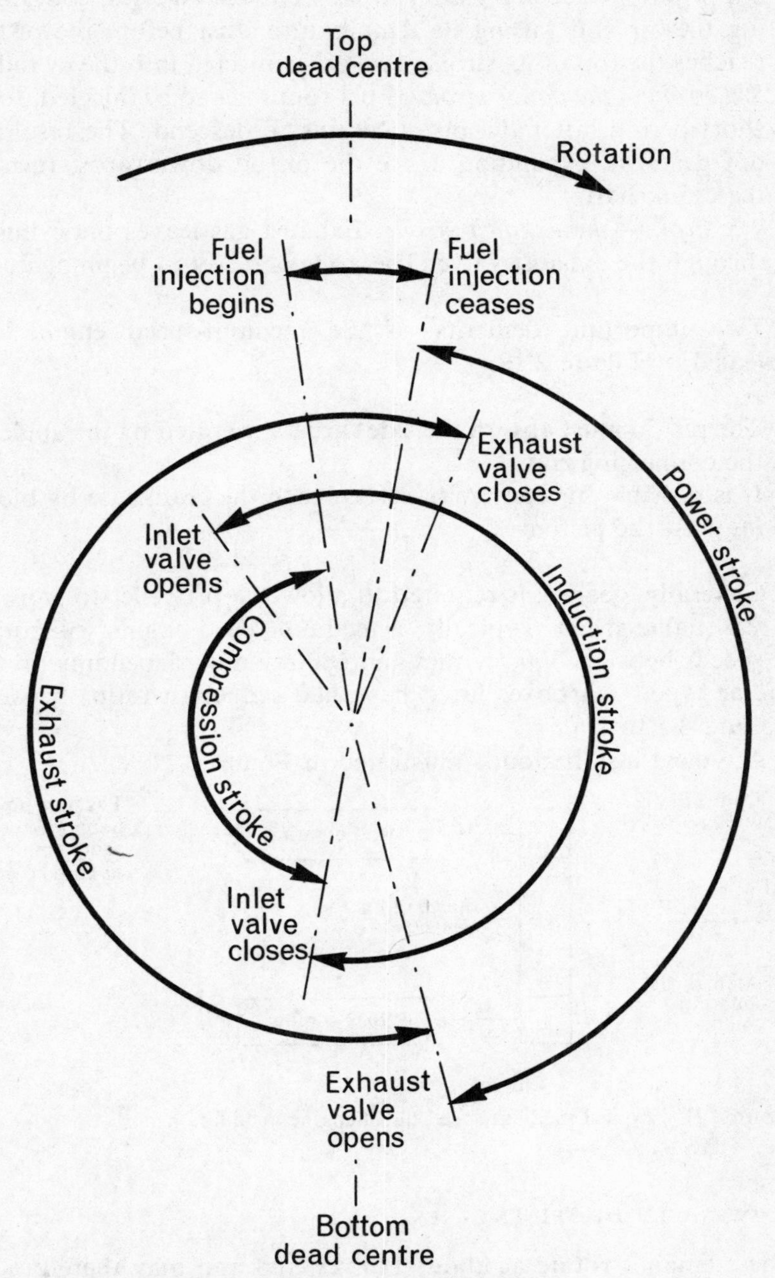

Figure 2.20 Timing of a four-stroke diesel engine.

2nd stroke—the compression stroke. The piston rises, compressing the air and raising its temperature. Just before the piston reaches the top of its stroke, fuel oil is injected into the cylinder.
3rd stroke—the power stroke. Fuel continues to be injected, for a short period, after the piston begins to descend. The resulting hot gases, in expanding, force the piston downwards, turning the crankshaft.
4th stroke—the exhaust stroke. Exhaust gas leaves the cylinder through the exhaust valve. The cycle is ready to begin again.

Two important demerits of the medium-speed engine are revealed by Figure 2.19:

1. The piston must absorb the side thrust generated by the angle of the connecting rod.
2. It is possible for contaminants to reach the crankcase by blowing past the piston.

Generally, gearing is required to allow the propeller to work at a reasonable speed. Typically, a medium-speed engine will rotate at speeds between 300 rev/min and 600 rev/min, depending on the engine type. Gearboxes fitted have had reduction ratios between 1·5 and 4·5 to 1.

A typical installation is illustrated in Figure 2.21.

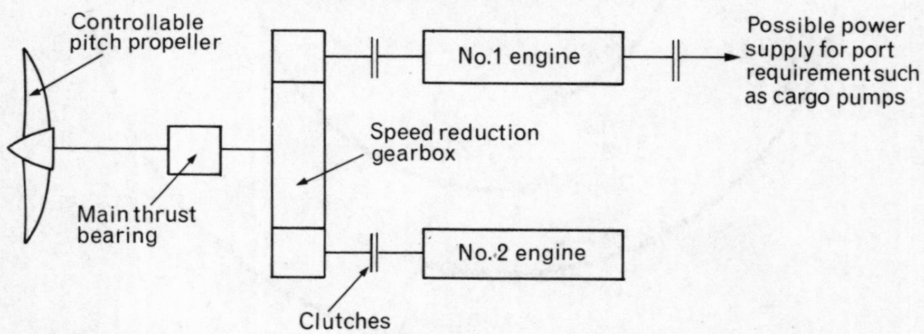

Figure 2.21 Typical medium-speed diesel-engine installation.

SLOW-SPEED DIESEL ENGINES

These engines rotate at about 110 rev/min and may therefore be directly coupled to a propeller. This type of engine is illustrated in Figure 2.22. It will be noted that the engine is taller than the

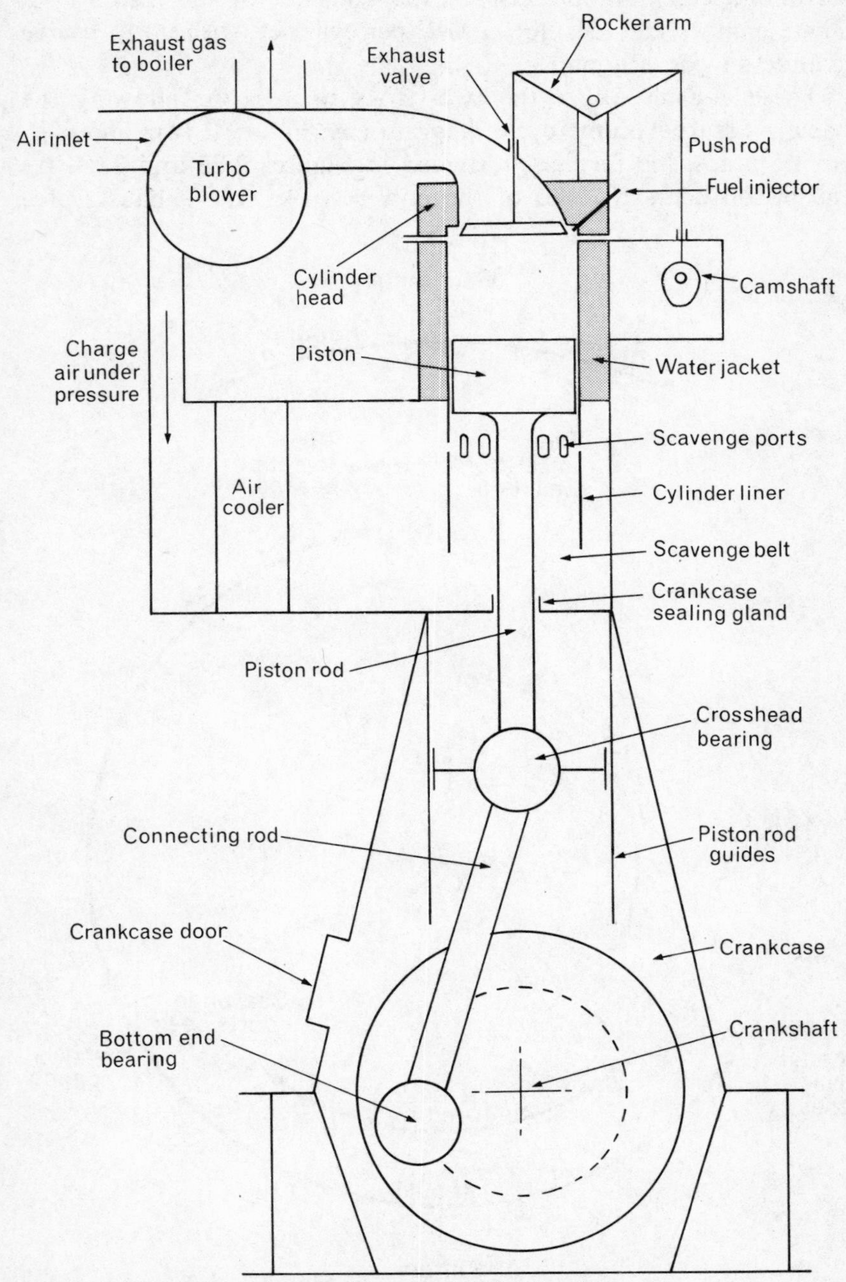

Figure 2.22 Typical slow-speed crosshead diesel engine.

medium-speed unit, because of the addition of the piston rod. These engines may develop 3 MW per cylinder, with up to twelve cylinders in one engine.

These engines utilize the two-stroke process in following the basic pressure–volume cycle diagram (*see* Figure 2.18). The two-stroke process is further illustrated in Figures 2.23 and 2.24. As the piston nears the end of the power stroke, the exhaust valve

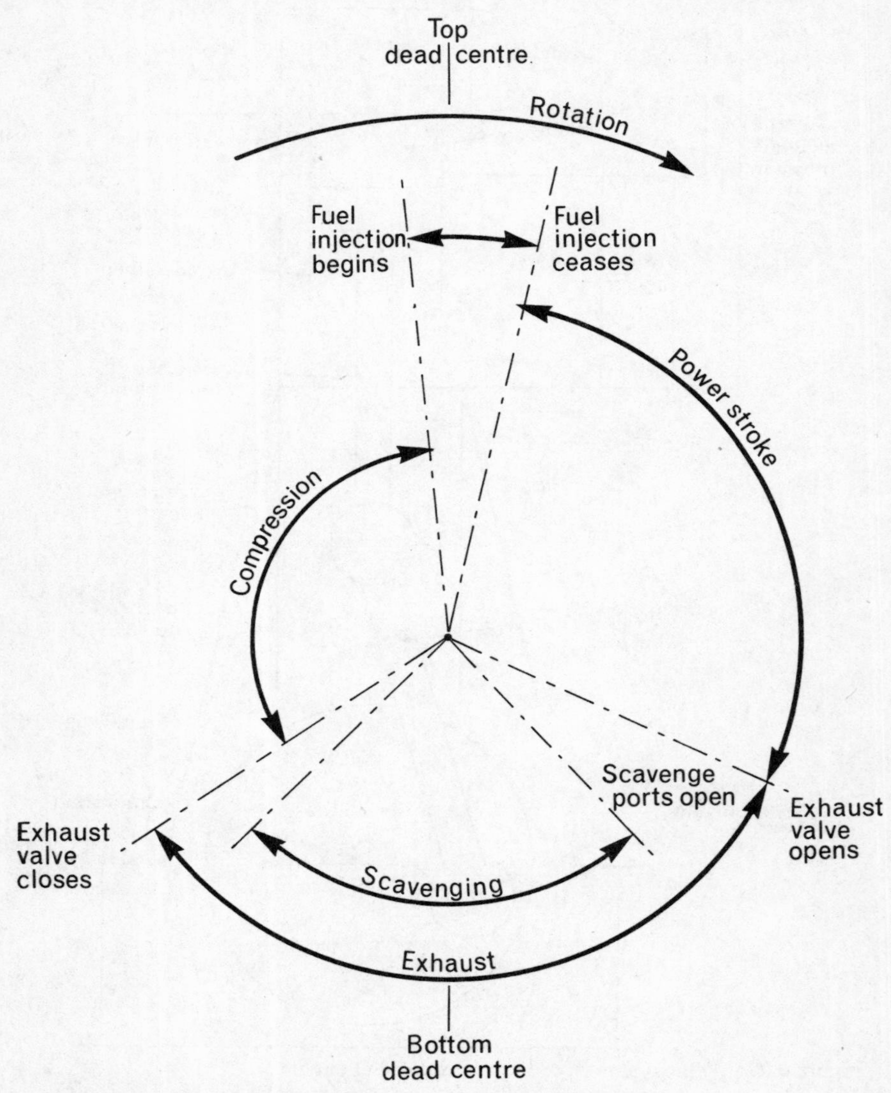

Figure 2.23 Timing of a two-stroke diesel engine.

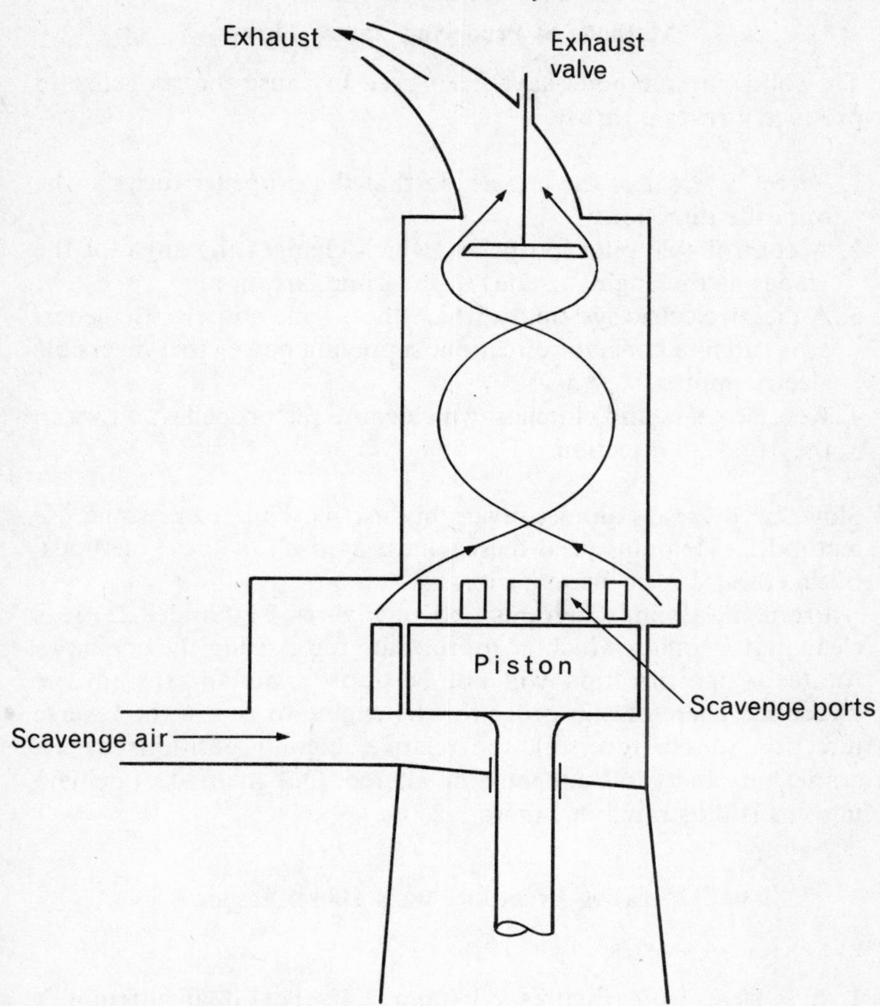

Figure 2.24 UNIFLOW scavenging process.

opens, allowing the pressure in the cylinder to fall rapidly as the gases escape from the cylinder. This enables scavenge air to enter the cylinder when the scavenge ports are uncovered by the descending piston. The cylinder is thus scavenged of exhaust gases, in this case by the UNIFLOW scavenging process, with scavenge air at a pressure of about 1·5 bar. The piston rises, covering the scavenge ports, the exhaust valve closes, and compression begins. The piston rises further, compressing the air to a high temperature, fuel is injected, and the cycle is repeated.

Methods of Producing Astern Thrust

The following methods have been used to cause the propeller to produce a reverse thrust.

1. Direct reversal of the engine, so that the propeller turns in the opposite direction.
2. A controllable pitch propeller, which changes the angle of the blades as the engine rotates in the same direction.
3. A diesel–electric system, in which the engine and electric generator run in a constant direction, supplying power to a reversible electric motor.
4. Reverse gears and clutches, which cause the propeller to turn in the opposite direction.

Slow-speed diesels almost invariably use method 1, but some use method 2. Medium-speed diesels have used all of these methods, but method 2 is in the majority.

From the timing diagrams above (Figures 2.20 and 2.23) it is clear that a timing which is appropriate for causing the engine to rotate in one direction will not be appropriate for the reverse direction. Therefore, in order for the engine to run in the reverse direction (direct reversal), the relative angular positions of the crankshaft and camshaft must be altered. One method of achieving this is illustrated in Figure 2.25.

Operating Procedure for a Diesel Engine

PREPARATIONS FOR STARTING

1. It is clear from Figures 2.19 and 2.22 that, if an attempt is made to start a diesel engine from cold, the piston will tend to expand at a greater rate than the cylinder liner, because of the higher thermal inertia of the liner and cooling water jacket. Therefore, before a large diesel engine is started the jacket water must be heated slowly, either by circulating hot water from generators or by blowing steam into the cooling water circuit.
2. All tanks, filters, and drains are checked. The lubricating oil and circulating pumps are started. All oil returns are checked.
3. All alarms and control equipment are checked.
4. Indicator cocks, which connect the space above the piston to

the atmosphere, are opened. The engine is then turned slowly with the turning gear, and any water which may have collected above the piston may be discharged through the indicator cocks.

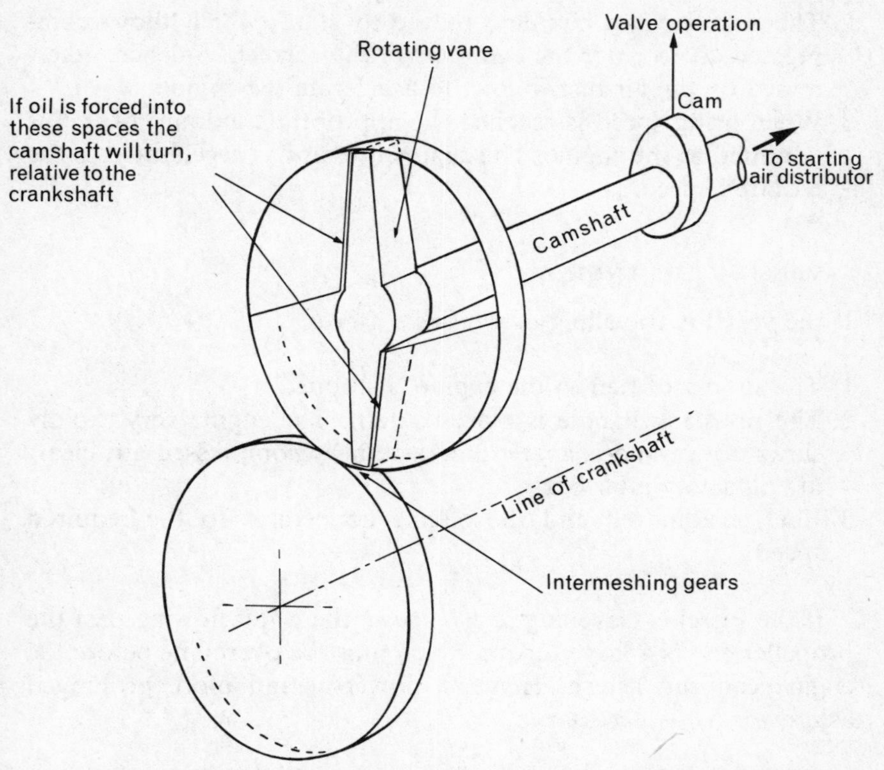

Figure 2.25 Method of altering the relative positions of the crankshaft and camshaft before reversing a diesel engine.

5. The fuel oil system is checked and warmed through. The air starting system is checked. The engine may require the starting of an auxiliary scavenge air blower.
6. The turning gear is removed and the indicator cocks are closed.
7. If possible, the engine should be kicked over by air.
8. The engine is now ready for starting.

Clearly, the length of this procedure depends on the size of the engine. A small diesel engine may be put on load in a very short time.

STARTING THE ENGINE

1. The reversing handle, which positions the camshaft relative to the crankshaft, is placed ahead or astern, which also places the distributor of starting air in the required position.
2. The manoeuvring handle is moved to 'start', which allows compressed air to enter the cylinders in the correct sequence, determined by the air distributor, to accelerate the engine.
3. When firing speed is reached the appropriate amount of fuel is admitted to the engine. The engine fires and is accelerated to the required speed.

REVERSING THE ENGINE

If the vessel is travelling *at moderate speed*:

1. The supply of fuel to the engine is stopped.
2. The reversing handle is placed astern. The engine very rapidly slows down, and is accelerated astern by compressed air, clearing all safety interlocks.
3. Fuel is admitted and the engine accelerates to the required speed.

If the vessel is travelling *at full speed*, the water flowing past the propeller exerts a large torque which must be overcome before the engine can run astern. Hence a powerful and more prolonged astern torque is necessary.

1. The fuel is shut off and the engine may require the starting of an auxiliary air blower.
2. Compressed air is applied to slow down the engine. This may be repeated at short intervals.
3. The engine slows and stops. The camshaft is repositioned and compressed air accelerates the engine to firing speed astern. Fuel is admitted, and the engine accelerates to the required astern speed.

Diagrams illustrating the stopping characteristics of the vessel under different conditions are usually available on board. Reference to these diagrams will help to clarify this last point and will lead to an appreciation of the time intervals involved.

GAS TURBINES

Gas turbines work on the same principle as steam turbines; i.e. a force is generated on turbine blades by changes in the velocity of the mass flow of fluid passing over the blades. The gas turbine uses the hot gases resulting from the combustion of fuel as its working fluid, as opposed to the steam turbine which uses the steam generated in boilers. A typical gas turbine plant is illustrated in Figure 2.26.

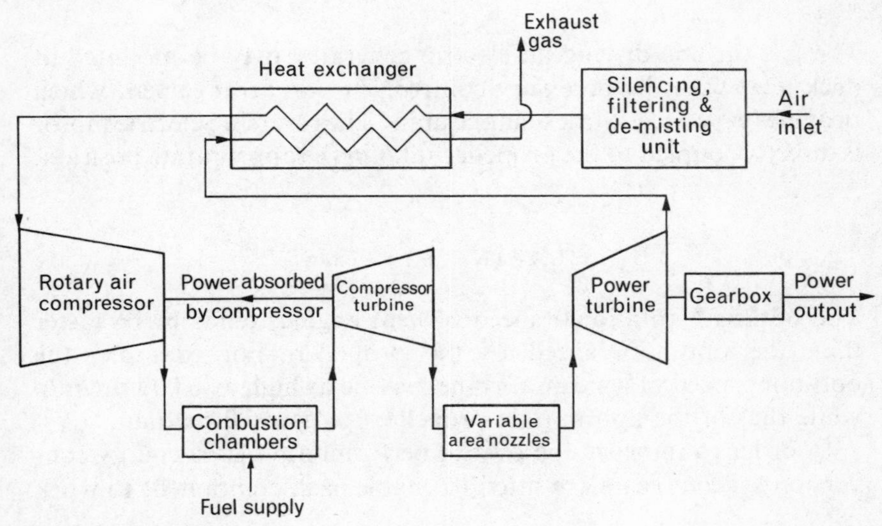

Figure 2.26 Gas turbine.

Air enters the system through a demister which removes salt-water spray from the air. The air is compressed in a rotary axial-flow compressor, then passes to the heat exchanger, which extracts energy from the exhaust gases and so improves the power plant efficiency. The compressed air now enters a number of combustion chambers, where fuel oil or liquefied gas boil-off may be burned. The resulting hot gases first pass through a turbine, the power output of which drives the compressor. The hot gases continue to expand through variable area nozzles, which control the output of the power turbine. The exhaust gases pass through the heat exchanger, and sometimes through an exhaust gas boiler, before passing to the atmosphere.

Gas turbines have the advantages of high power to weight ratios and subcomponents which may be changed very quickly. However, their energy conversion efficiency is lower than that of some other power plants, and the fuel oil that they use must usually be specially treated.

Most gas turbines are not directly reversible; hence reverse propeller thrust must be obtained by one of the following:

1. A controllable pitch propeller.
2. Reversing clutches and gears.
3. A turbo–electric system.

The gas turbine driving an electric generator may be mounted at deck level with the necessary control gear, an arrangement which provides easy access for exchange units. The reversing electric motor is directly coupled to the propeller shaft in the appropriate position.

GEARING SYSTEMS

The optimum rotational speed of heat engines tends to be faster than the optimum speed of the propeller. For example, the optimum speed of a steam turbine may be as high as 6 000 rev/min while that of the appropriate propeller is about 70 rev/min.

In order to improve the overall performance of the energy conversion system, gears are fitted to enable each component to work

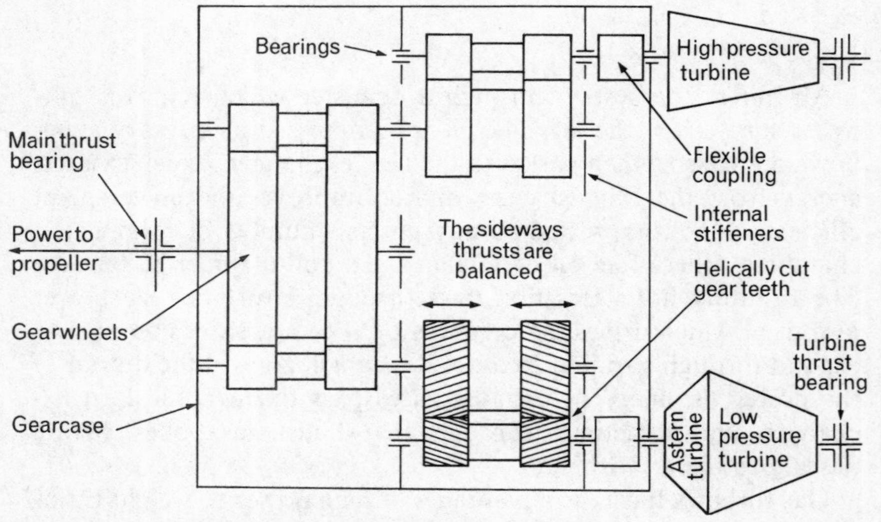

Figure 2.27 Turbines with double reduction gearing.

efficiently. This is especially true for steam and gas turbines and for medium-speed diesel engines. Figure 2.27 illustrates a double reduction gear system for a steam turbine. The gear teeth are out on a helix, like a screw thread, so that the force between mating gears is distributed over several teeth. The helix tends to make the gears slide sideways; hence double gear wheels are used, with the gears on each wheel cut in opposite directions, to cancel out the sideways force.

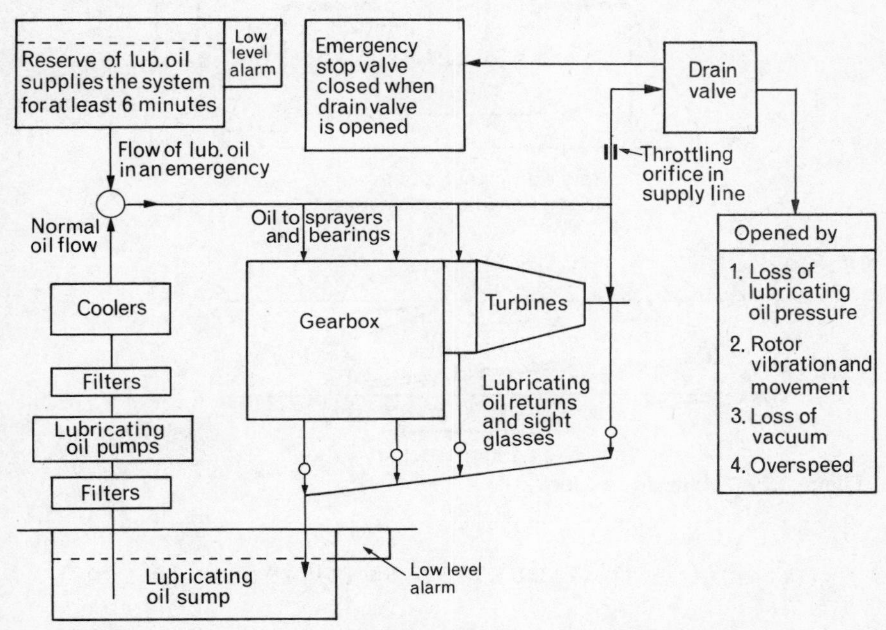

Figure 2.28 Lubrication system for a steam turbine installation.

Figure 2.28 illustrates a lubrication system for a steam turbine installation. The lubricating oil pressure is used as a control medium to shut down the power plant by closing the emergency stop valve, for various developing conditions which are listed in Chapter 4 in Figure 4.7 and under the heading 'Bridge Control of Steam Turbines'.

The thrust block (*see* Figure 2.29) absorbs the thrust force created by the rotating propeller and transmits this force to the ship structure. The thrust collar is forged as part of the propeller shaft, and tilting pad bearings absorb the transmitted force, transferring the force to the thrust-bearing housing and hence to the ship structure.

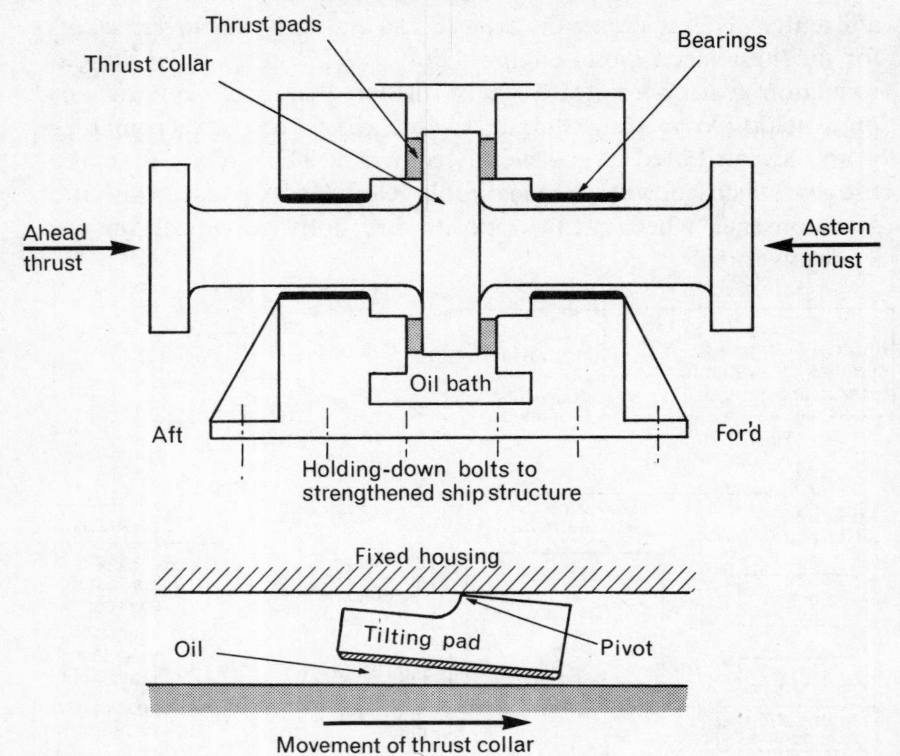

Figure 2.29 Main thrust block.

DISTILLATION SYSTEMS

PROBLEM OF SCALE FORMATION

Sea water contains various salts which together form about one thirty-second of the mass of sea water. The salts are present in the following approximate proportions:

Sodium chloride	80%
Magnesium chloride	10%
Magnesium sulphate	6%
Calcium sulphate	4%

If this sea water were used in the water-tube boiler described above, it would leave behind on evaporation:

$$(1/32) \times 80 \times 10^3 = 2{\cdot}5 \times 10^3 \text{ kg of salts per hour.}$$

These salts would tend to form scale on the heat transfer surfaces of the boiler, which would result in overheating of the tubes, combustion chambers, and furnaces which form the heat transfer surfaces. Scale formation and its consequences are illustrated in Figures 2.30 and 2.31.

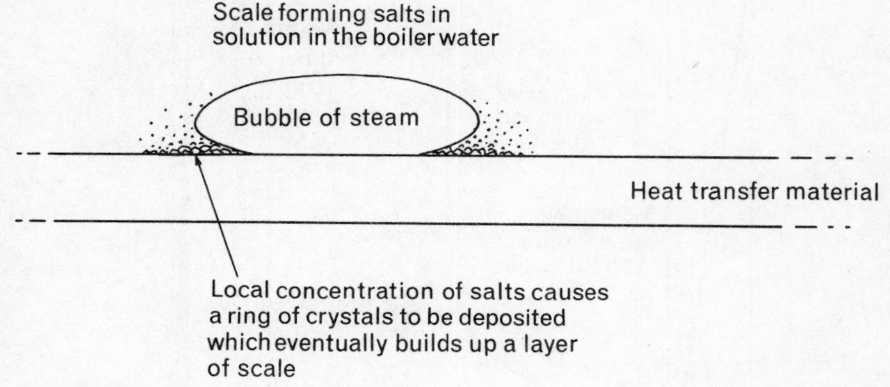

Figure 2.30 Scale formation.

Clearly, this process must be prevented, to allow for the safe operation of the boiler and to maintain the availability of the vessel for revenue earning. To prevent scale formation, four techniques are possible:

1. Keep a reserve of chemicals in the boiler, which will combine with any potential scale-forming salts entering the boiler and effectively neutralize them.
2. Pass all feed water through units in the feed system, which will convert scale-forming salts into a harmless form.
3. Ensure that scale-forming salts do not enter the boiler in any significant quantity, by using an effective distillation system.
4. Use a combination of all three methods, which is the course often followed in water-tube boiler installations.

DISTILLATION SYSTEMS

A distillation system is installed to provide pure water from sea water, to be used for both domestic and engineering purposes. Distilled water is obtained by causing sea water to evaporate by either a boiling process or a flash process, which reduces the

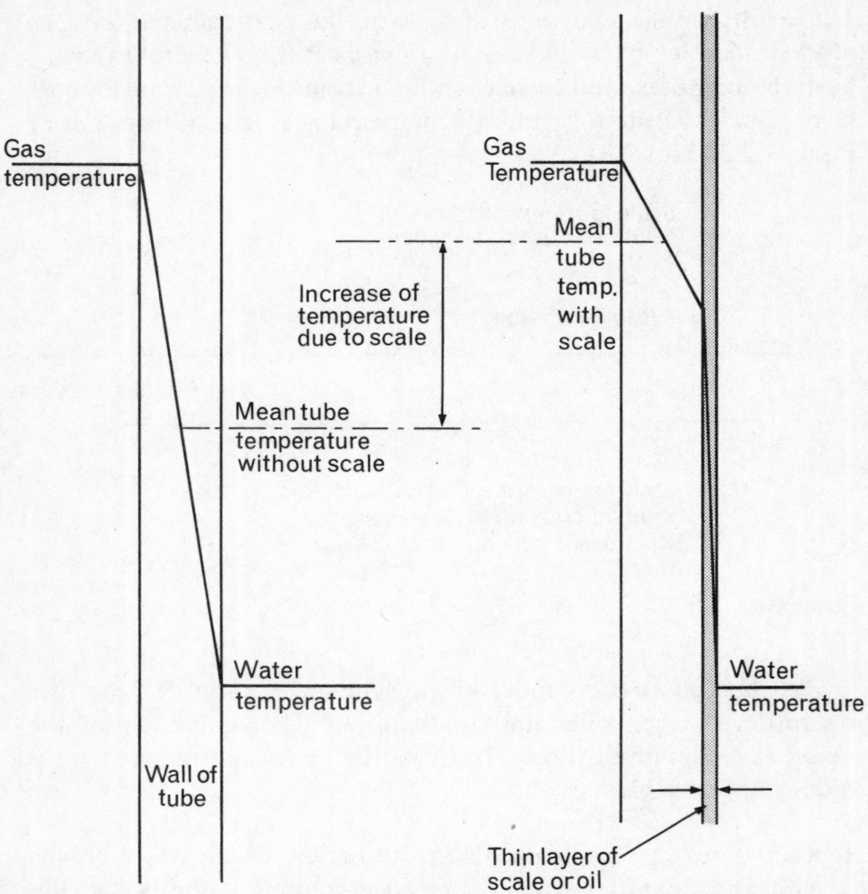

Figure 2.31 Effect of scale or oil on the temperature of a heat transfer surface.

31 250 parts per million of salts in sea water to about 2 parts per million in the distillate.

Figure 2.32 illustrates a plant using a *boiling process*. Water is boiled by the energy supplied by the heating coil, under low-pressure conditions, the boiling temperature being about 60 °C. Steam rises from the boiling sea water, leaving behind its associated salts, which are then pumped overboard dissolved in the brine. The rising steam passes through a separator, which removes any fine droplets of sea water suspended in the steam, and then it passes over tubes containing sea water, which results in the steam's condensing to form distilled water of high purity.

Figure 2.33 illustrates a plant using the *flash process*. Sea water

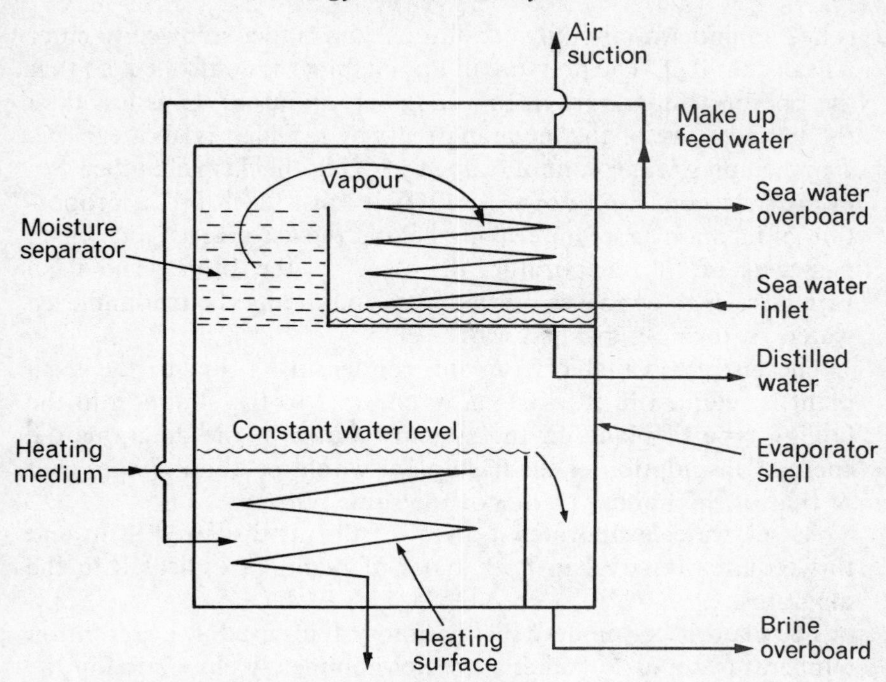

Figure 2.32 Distillation plant using a boiling process.

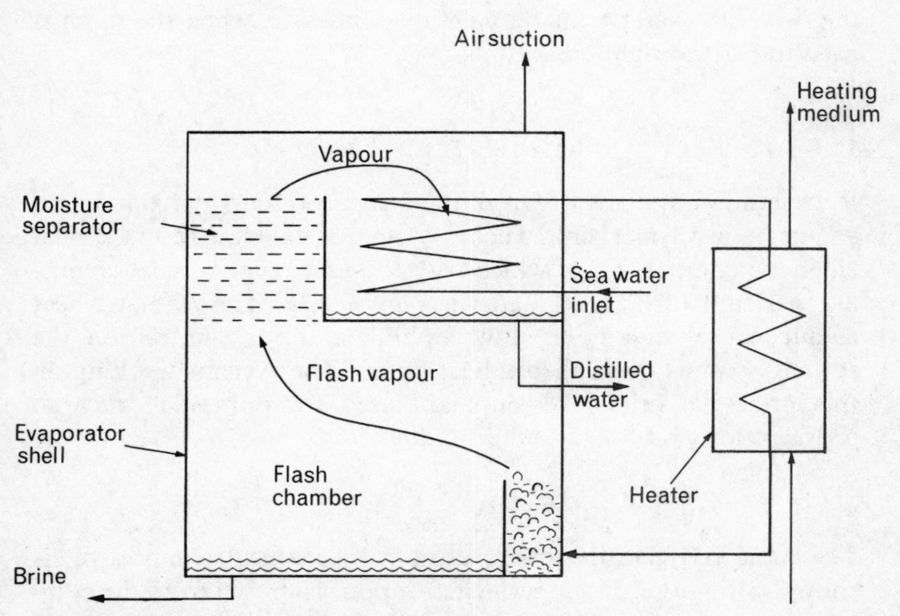

Figure 2.33 Distillation plant using the flash process.

is heated under nonboiling conditions, and is then allowed to enter a flash chamber. The pressure in this chamber is controlled, so that the boiling temperature in this chamber, about 40 °C, is less than the temperature of the incoming sea water, which is about 65 °C. The incoming water cannot exist at 65 °C in the flash chamber. So, in order to cool itself down to 40 °C it must 'flash off' a proportion of its mass as steam, leaving behind the associated salts which pass out of the evaporator, dissolved in the brine. The flash vapour rises to condense on the tubes containing the incoming sea water, so forming distilled water.

The energy available from the condensing steam in the flash plant is retained in the system, in contrast to the situation in the boiling type of plant. In the systems shown in the diagrams the energy consumption of the flash plant would be about 70 per cent of that of the boiling plant, for the same output.

As sea water evaporates it gives off the air dissolved in it, and this requires removal in both types of plant, as indicated in the diagrams.

The majority of modern distillation systems use low evaporation temperatures and so bacteria are not completely eliminated in the evaporation process. Care must therefore be taken not to use these plants to produce water for domestic purposes while the ship is in polluted water. Chemical treatment plants are fitted to ensure that the domestic water is made palatable and safe, when the plant is used under the right conditions.

REFRIGERATION SYSTEMS

Refrigeration systems organize the transfer of energy from a cold source to a warmer sink. Such a system is called a reversed heat engine, a comparison of which with the usual type of heat engine, e.g. a steam turbine, is made in Figure 2.34. The reversed heat engine causes energy to flow 'uphill', and for this reason the system is sometimes called a heat pump. The systems working on this principle, on board ship, are used for domestic or cargo refrigeration and for air conditioning.

BASIC REFRIGERATION CYCLE

The basic refrigeration cycle, using a mechanical compressor, is shown in Figure 2.35. Superheated vapour is drawn from the com-

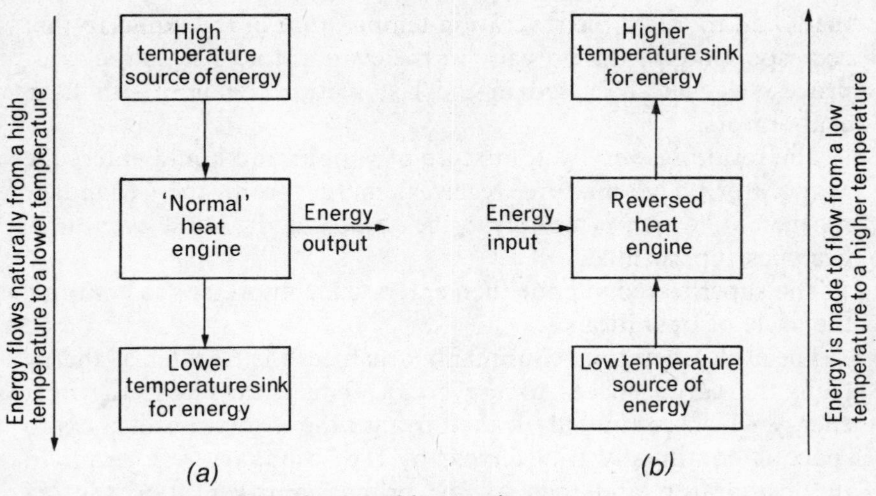

Figure 2.34 'Normal' and reversed heat engines.

pressor and discharged to the condenser. The superheated vapour entering the condenser first has its superheat removed and is then condensed. The resulting liquid is then cooled to just above the temperature of the incoming sea water.

This liquid then flows through the expansion or regulating valve. As soon as the liquid enters the region of lower pressure a proportion evaporates. Evaporation continues until sufficient latent heat of evaporation has been taken from the remainder of

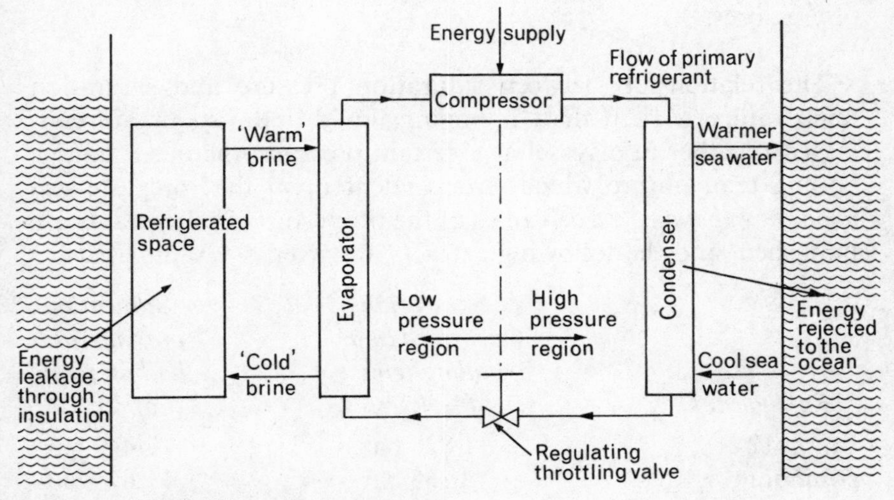

Figure 2.35 Refrigeration system.

the liquid to lower the saturation temperature of the liquid to that corresponding to the pressure in the evaporator. This is the same process as was discussed in the last section, dealing with flash evaporators.

The resulting, very wet, mixture of vapour and liquid enters the evaporator. The mixture receives energy from the secondary refrigerant, which is usually brine, becomes dry, and eventually becomes superheated.

The superheated vapour then enters the compressor to continue the cycle of operations.

The cycle, therefore, continually organizes the transfer of energy from the cargo spaces to the ocean. This means that the heat energy which continually leaks through the insulation of a cargo space is continually transferred, by the secondary refrigerant to the evaporator, and then by the primary refrigerant to the sea water circulating in the condenser.

PROPERTIES OF PRIMARY REFRIGERANTS

When the refrigeration cycle is examined it becomes clear that, for a refrigerant, a fluid is required which will:

1. Boil at conveniently low temperatures at reasonable pressures.
2. Condense at about normal sea-water temperature at reasonable pressures.

The relationship between saturation pressure and saturation temperature is such that, if boiling liquid and vapour are contained together in a vessel at a certain pressure, there is a unique boiling temperature which is dependent upon that pressure *and upon the substance used.* Tables of the properties of refrigerants are published, and the following extracts are given as examples.

Refrigerant	*Saturation pressure at a boiling temp. of −15 °C*	*Saturation pressure at a boiling temp. of 30 °C*
Freon 12	1·826 bar	7·449 bar
Ammonia	2·365 bar	11·67 bar
Carbon dioxide	23·400 bar	72·80 bar

Water has been used in some very large air-conditioning plants, but the lowest boiling temperatures available are relatively high—too high for refrigeration plants.

SECONDARY REFRIGERANTS

Secondary refrigerants are used to transfer energy from the cargo spaces to the refrigeration plant. They are not used in small plants. A bar refrigerator does not require this facility; the cooling coils in the refrigerated space contain the primary refrigerant. Secondary refrigerants are usually brine or air.

1. *Brine systems.* Brine is a solution of calcium chloride in fresh water; as such it is basically noncorrosive and cheap. Salt water should not be used to mix brine, as this solution will promote corrosion. The brine is drawn from the evaporator by a pump and delivered to grids of pipes in the cargo spaces, which absorb heat from the air in the spaces. The brine then returns to the evaporator, and the temperature of the brine returns are monitored, as are the cargo spaces.
2. *Air cooling systems* are used especially for the carriage of fruit and meat. Brine is circulated through heat exchangers, and a fan blows air across them, circulating it through the cargo spaces. Facilities are fitted to allow the air in fruit spaces to be refreshed, in order to reduce the carbon dioxide content.

PUMPS AND PUMPING SYSTEMS

A pump is a device which transfers energy to a fluid passing through it. There are many forms of energy, but when pumps are being considered, use can be made of the energy equation illustrated in Figure 2.36. The *pump efficiency* can be defined thus:

$$\text{Pump efficiency} = \frac{\text{Rate of energy transfer to fluid}}{\text{Energy flow supplied to the pump}}$$

In practice, pumps change both the velocity and the pressure of the fluid passing through them, the ratio of these energy changes depending on the type of pump being used.

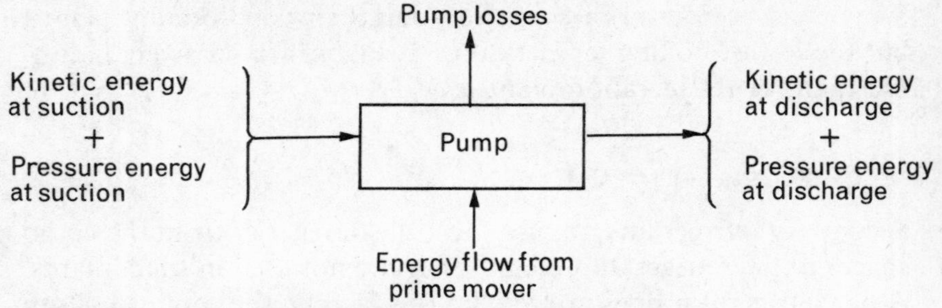

Figure 2.36 Major energy flows through a pumping system.

There are three main classes of pumps: positive displacement, axial flow, and centrifugal.

Positive Displacement Pumps

RECIPROCATING PUMPS

A line diagram of a double-acting pump of this type is shown in Figure 2.37. The discharge pressure variations of this sort of pump are illustrated in Figures 2.38 and 2.39. It will clearly be seen that the function of the air vessel is to reduce the pressure fluctuations which result from the harmonic motion of the pump piston, as illustrated by Figure 2.38. The pressure–flow characteristic shows

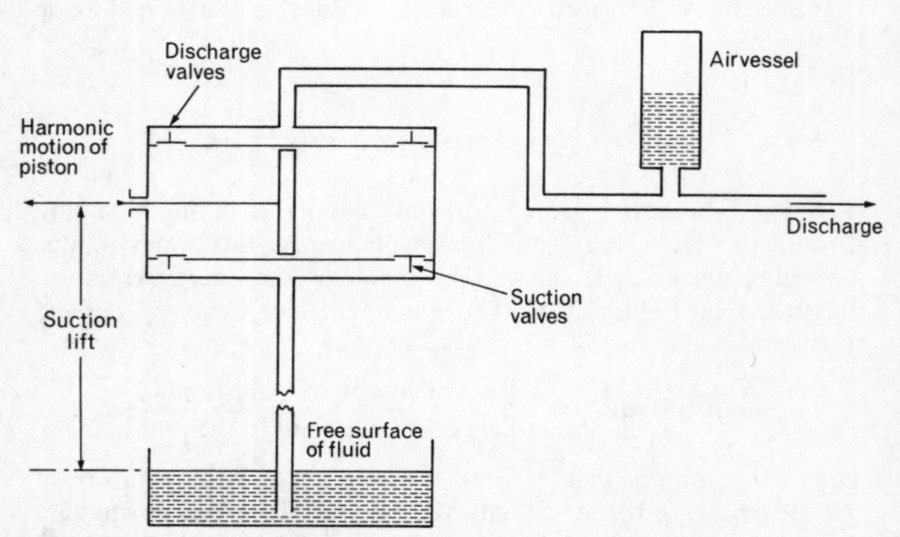

Figure 2.37 Double-acting reciprocating pump.

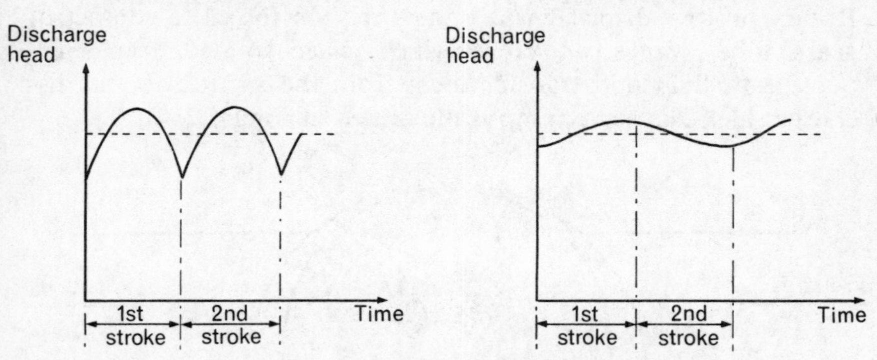

Figure 2.38 Fluctuation in discharge pressure with and without air vessel.

that rate of flow, at any particular speed, is almost independent of discharge pressure.

The advantages of reciprocating pumps include the ability to handle large proportions of vapour, which enables them to deal with volatile or hot liquids. Also, they are self-priming and can handle high-suction lifts. However, their construction is complicated by the presence of suction valves, discharge valves, and air vessels and by the need for relief valves to protect the pump against the closure of delivery lines.

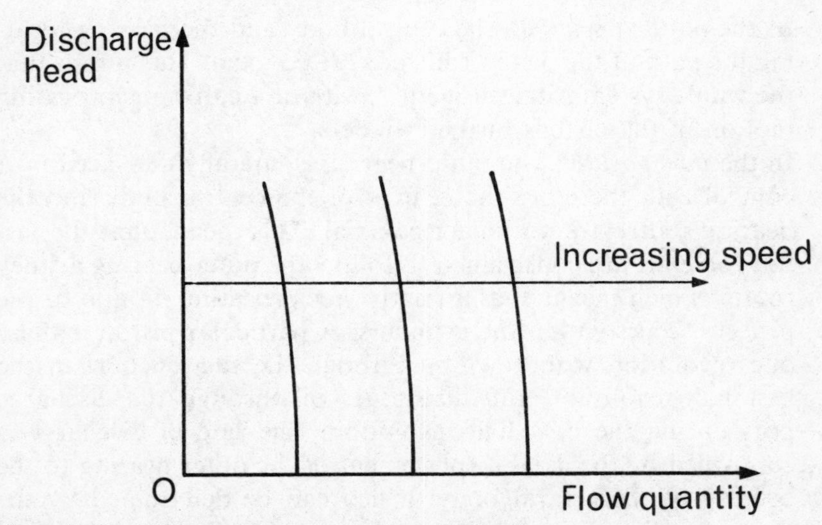

Figure 2.39 Characteristics of a positive displacement pump.

ROTARY PUMPS

Rotary positive-displacement pumps rely on the close contact of gears, lobes, vanes, or screws which, when rotated, trap small pockets of fluid and transfer them from the suction to the discharge side. One such pump is illustrated in Figure 2.40.

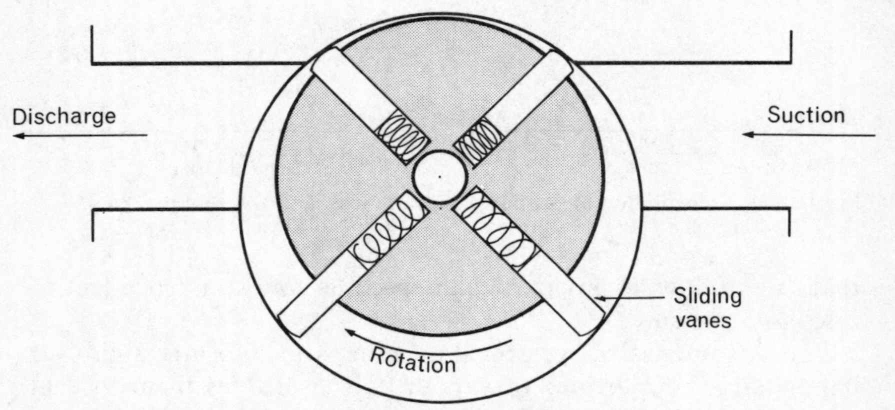

Figure 2.40 Rotary positive-displacement pump.

Constant-speed variable-flow rotary pumps are used as power units in hydraulic control systems, e.g. ship steering gears. Figure 2.41 illustrates such a pump.

1. In the *no-flow position*, the central body and the bearing locating the path of the piston rod ends are co-axial. This means that the whole system rotates together, with no relative reciprocating motion of the pistons in the cylinders.
2. In the *flow position*, the outer bearing is laterally displaced by a control rod; therefore the centres of the central body and the bearing centre are no longer co-axial. This means that the piston rod ends are constrained to follow the outer bearing as they rotate, which means that a relative reciprocating motion of the pistons occurs within the cylinders. A particular piston making one revolution withdraws oil through the suction port in the first half-revolution and discharges oil through the discharge port during the next half-revolution. The rate of discharge is controlled by the axial displacement of the outer bearing to the bearing of the central body, which can be determined by the control rod.

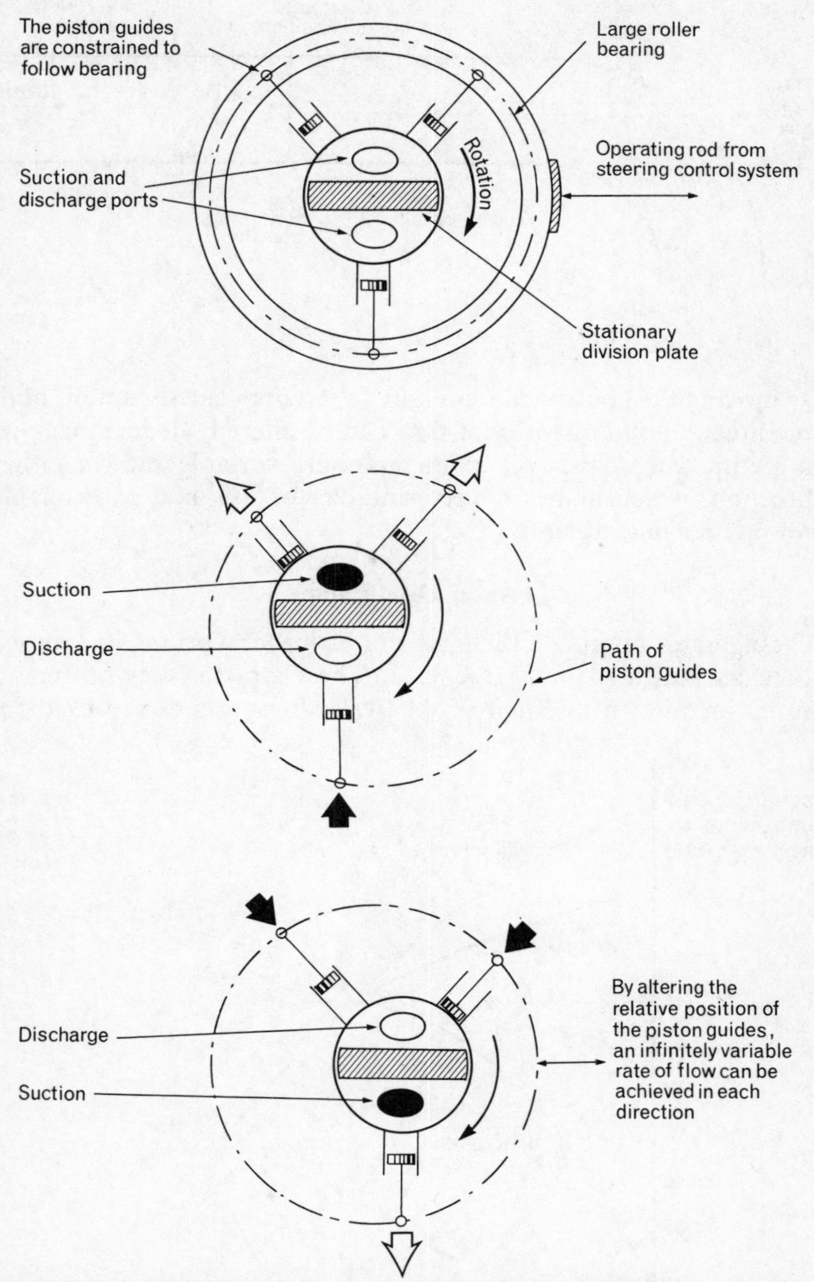

Figure 2.41 Operating principle of the Hele Shaw variable-delivery positive-displacement pump.

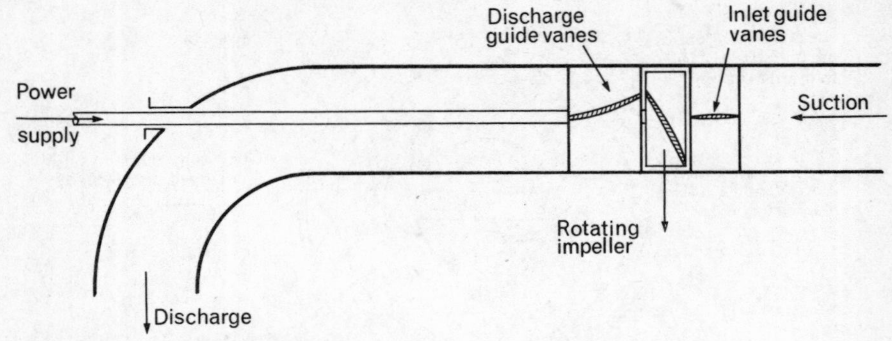

Figure 2.42 Axial flow pump.

By moving the control rod laterally in the opposite direction, both the direction and quantity of flow can be altered. Hence, this type of pump can deliver oil at an infinitely variable rate in either direction, which makes it very suitable for use as a controllable power-transmitting unit.

Axial Flow Pumps

These pumps approach the idea of a propeller working in a closed duct, as indicated in Figure 2.42. The characteristics of such a pump are shown in Figure 2.43. Axial flow pumps are used for

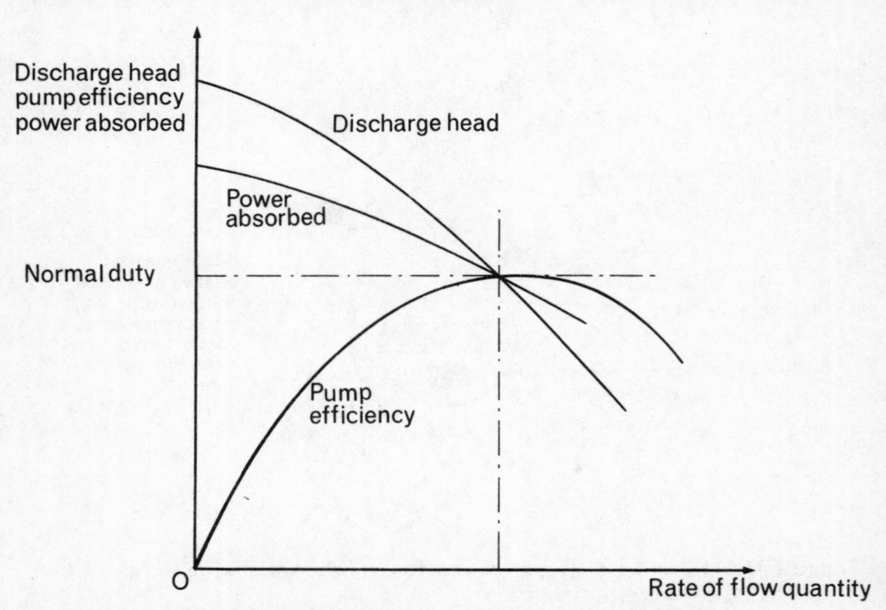

Figure 2.43 Performance characteristics of an axial flow pump.

applications which require a large flow rate against a relatively low discharge head, e.g. the circulating pump for a main condenser.

Centrifugal Pumps

Centrifugal pumps are distinguished from positive displacement pumps by their requirement of relative velocity between the fluid and the impeller. Figure 2.44 illustrates such a pump.

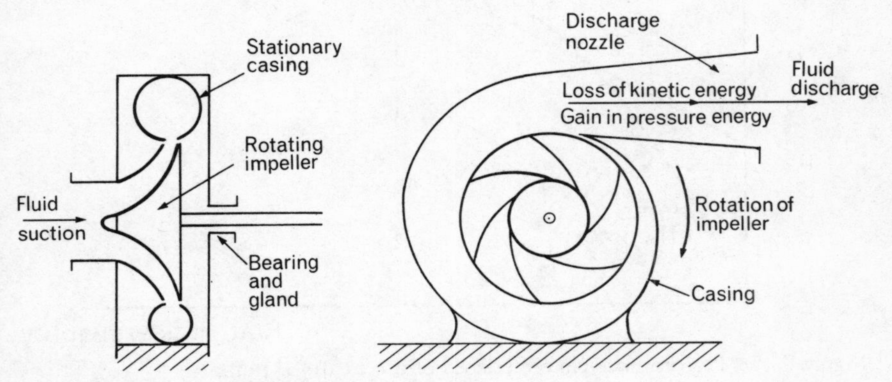

Figure 2.44 Centrifugal pump.

Fluid enters the 'eye' of the impeller and flows radially outwards through passages in the impeller, gradually increasing its linear velocity because of the impeller's rotation.

The fluid leaves the impeller in a similar manner to sparks shooting from a catherine wheel. The high-velocity fluid is collected in a specially shaped casing, where some of the kinetic energy of the fluid is converted into pressure energy. Should further conversion be required, the fluid is passed to a diverging discharge nozzle. As the fluid slows in this nozzle it loses kinetic energy which, according to the energy equation, must be replaced by a corresponding increase in potential energy.

The characteristic curves of a typical centrifugal pump are shown in Figure 2.45.

Suction Lift

Applying to pumps a principle learned in physics:

$$\text{Maximum possible suction lift} = \frac{\text{Pressure acting on free surface}}{\text{Specific weight of fluid}}$$

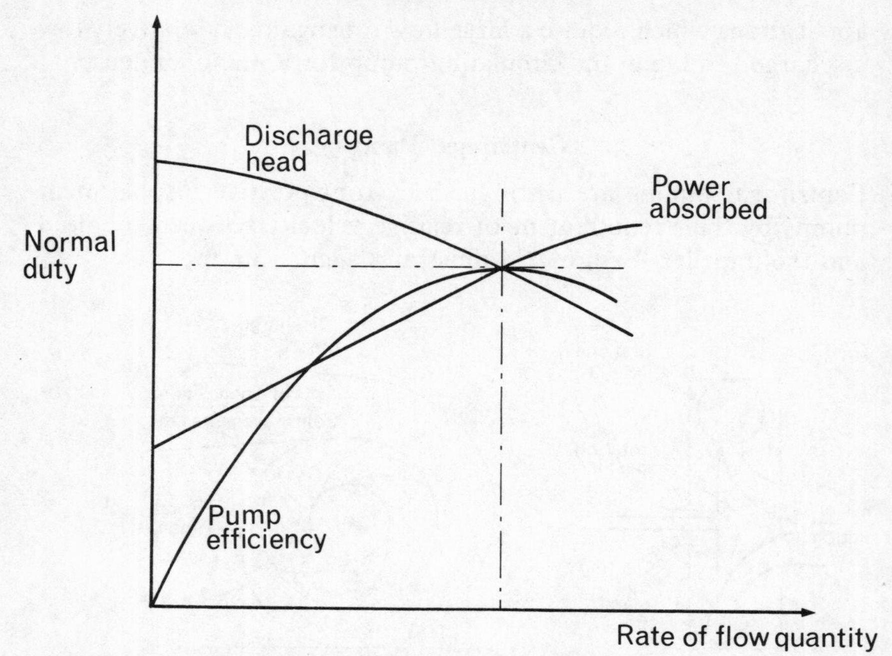

Figure 2.45 Performance characteristics of a centrifugal pump.

If the atmospheric pressure acting on the surface of fresh water is considered, it is found that the theoretical maximum lift works out to 10·36 m, which is the height of the column of fresh water which can be supported by normal atmospheric pressure. By the same argument it can be seen that the maximum suction lift of mercury would be 0·76 m.

In practice, these theoretical maximum figures are very much affected by three factors:

1. The temperature and volatility of the fluid which is being pumped. As the liquid approaches its boiling temperature, under the reduced pressures in the suction pipe, vapour is given off which reduces the suction exerted by the pump.
2. The pressure exerted on the free surface of the fluid. Clearly, special problems arise when fluid is being pumped from low-pressure spaces, e.g. condensers. These have to be solved by placing the pump at a level lower than that of the free surface of the fluid, which results in a positive suction head. The same problem is solved in gas carriers by submersing the impeller at the bottom of the cargo tank. Such pumps are called deepwell

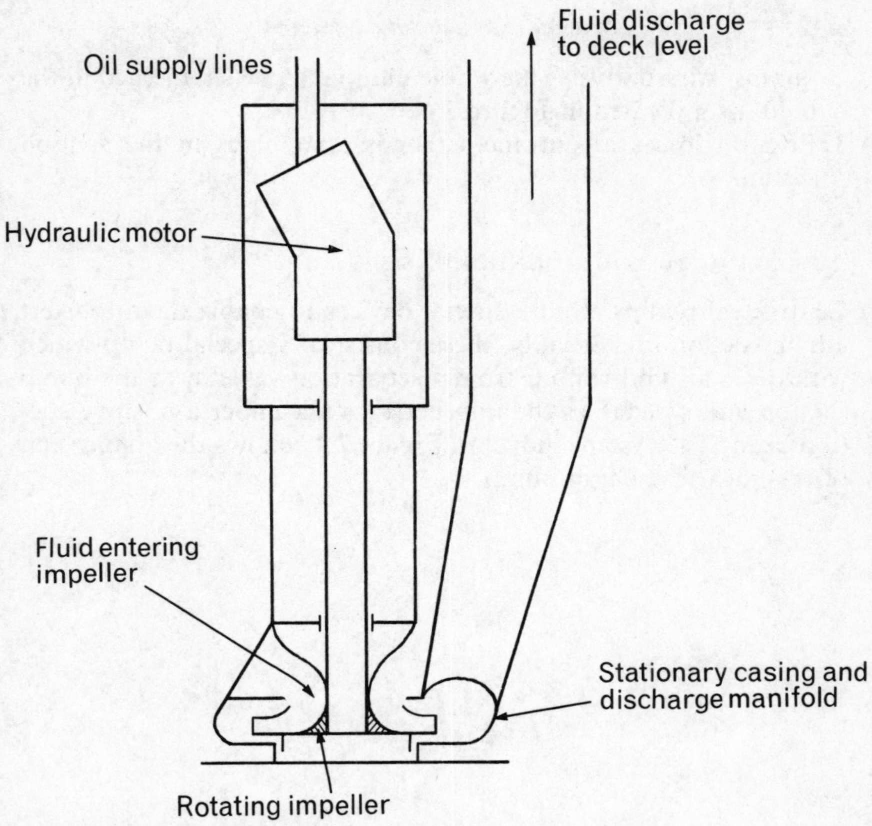

Figure 2.46 Hydraulically driven, submerged cargo-pump.

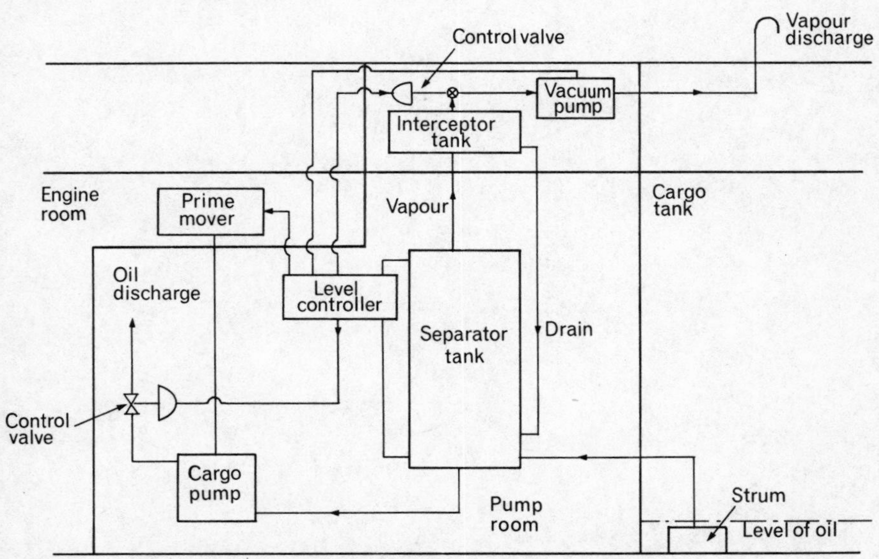

Figure 2.47 Stripping system using the main cargo pump.

pumps. Alternatively, the whole pump may be submerged in the fluid, as indicated in Figure 2.46.

3. Friction losses at entrances, bends, and pipes in the suction system.

IMPROVING SUCTION CONDITIONS

Centrifugal pumps require special devices to enable them to exert a high suction lift. Broadly, these consist of a special pump which withdraws air and vapour from a separation vessel near the pump suction and so enables the impeller to work under a positive suction head. The system shown in Figure 2.47 shows the application of this idea to a cargo pump.

Chapter 3

The Principles of Measurement, and Control Systems

Through the 1960s there was a rapid increase in the installation of complex centralized measurement systems and control systems on board ships. The most extensive application has been made in power plant systems, which has resulted in the unmanned machinery space (U.M.S.) concept's being widely applied. In parallel with this progress in power plant control there has been the development of centralized cargo control in oil, chemical, and liquefied gas carriers, all of which have required the development of the appropriate instrumentation.

CLOSED LOOP SYSTEMS

Shipboard systems can be examined under two headings: automatic and nonautomatic closed-loop systems.

NONAUTOMATIC CLOSED-LOOP SYSTEMS

These are characterized by the fact that no information is *automatically* fed back from the process under control to the regulating device which is controlling the process. Figure 3.1 illustrates the control of the level in a boiler, which is the controlled process, the human operator being necessary to 'close the loop'.

Other manually controlled processes on board can easily be recognized. For example, if an oil tanker is discharging its cargo,

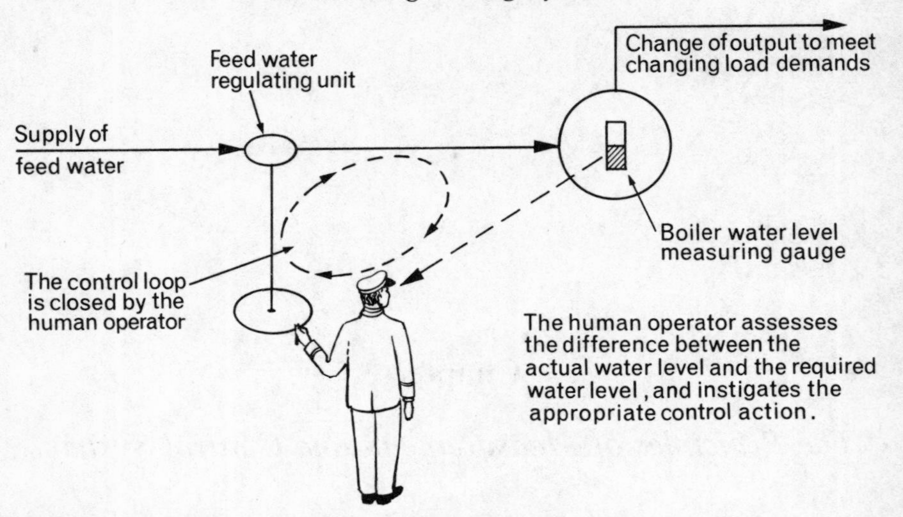

Figure 3.1 Example of a nonautomatic control system.

the person in the cargo control room receives information on tank levels, rate of flow, etc., processes this information in his brain, and manually operates the regulating device controlling the discharge process as necessary.

AUTOMATIC CLOSED-LOOP SYSTEMS

These are characterized by the fact that information about the performance of the system under control is automatically fed back to a controller, which instigates the proper corrective action.

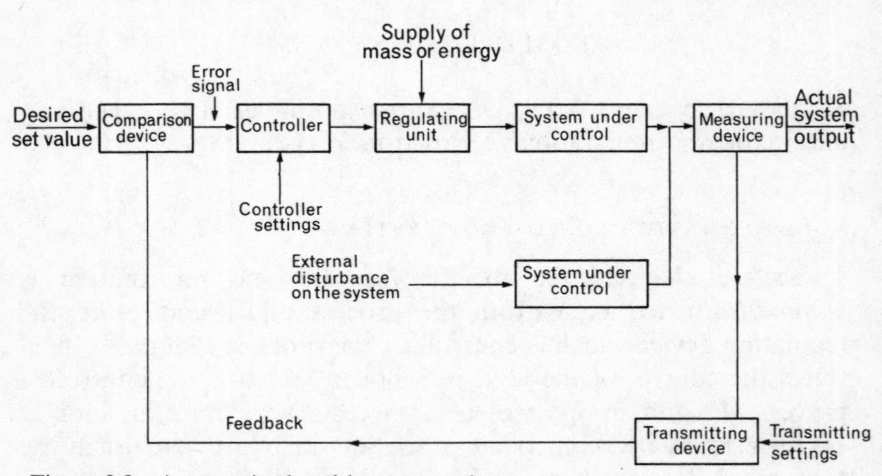

Figure 3.2 Automatic closed-loop control system.

Automatic closed-loop control is illustrated in Figure 3.2. The output from the system under control, i.e. the controlled variable, is measured, and this information is transmitted to a comparison device, which compares the actual value of the measured variable with the desired value. The output of the comparison device is a function of any difference between the actual and desired values, which is usually called the error. The signal from the comparison unit is fed to the controller. The controller manipulates the error signal and sends a modified signal to the regulating unit, which controls the supply of mass or energy to the system. The system output, which is the controlled variable, eventually responds to the change in input, so that the controlled variable approaches the desired value. The response is detected by the measuring unit.

If this block diagram is applied to the previous example, it can be seen how a measuring device might continually measure the boiler water level, transmitting this information to the comparison device, which would compare the measured water level with that required value set by the engineer officer. The signal from the comparison unit, modified by the controller, would pass to the regulating unit, which would open or close the valve controlling the supply of feed water to the boiler. Eventually the level in the boiler would respond to the change in input, the response being detected by the measuring unit.

If an external disturbance is imposed on the boiler, e.g. during manoeuvring, the change of load is reflected in a change in water level, which is detected by the measuring unit, which in due course causes the regulating unit to adapt the flow of water into the boiler to meet the changing load requirements.

The validity of replacing the monotonous exercise of controlling boiler water level by hand, with some kind of automatic closed-loop control, would appear to be immediately apparent.

Clearly, before a certain system output variable can be controlled, it must first be measured. Measuring devices are extremely varied and often very diverse physical principles may be used to measure any particular variable. The measurement of certain variables especially appropriate to shipboard practice will now be considered.

MEASUREMENT OF PRESSURE

(*a*) BOURDON TUBE

Because of its reliability and versatility, many pressure-measuring devices use a Bourdon tube, shown in Figure 3.3. An increase in the internal pressure, or a decrease in the external pressure, tends to make the section of the tube become circular, which tends to straighten the tube. The resulting movement can be used in many different ways; e.g. the diagram shows how a pointer may be made to move over an indicating scale.

(*b*) BELLOWS ELEMENT

Bellows are made by hydraulically forming a tube in a die. When made they have a certain flexibility; hence a change in external or

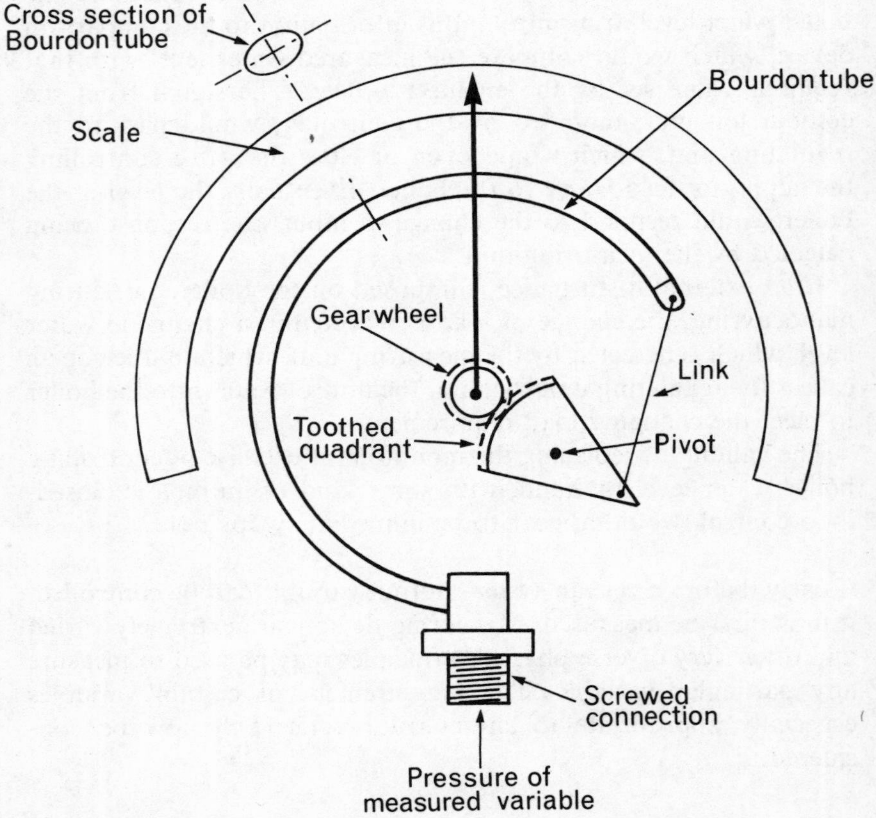

Figure 3.3 Bourdon pressure gauge.

internal pressure causes the bellows to flex and so allows a change of pressure to be measured, as indicated in Figure 3.4. The bellows unit is extremely versatile and is used in many components of control systems.

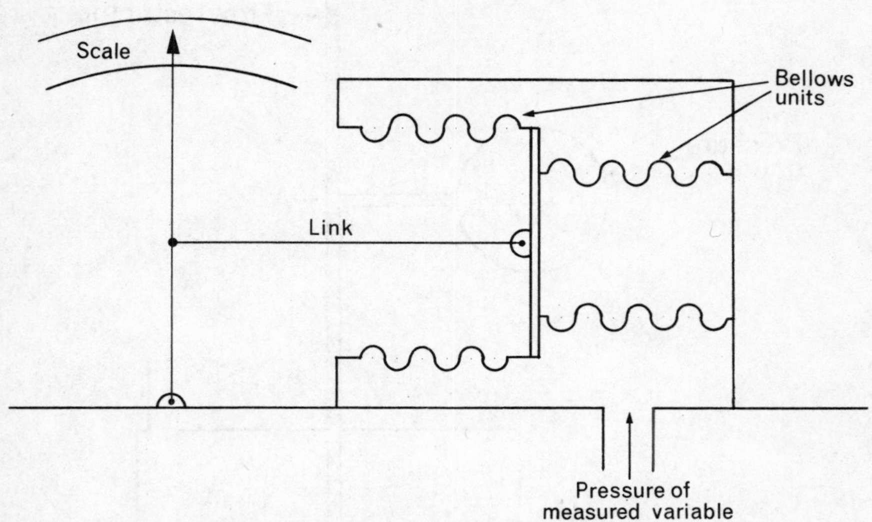

Figure 3.4 Pressure measurement using bellows units.

MEASUREMENT OF LEVEL

On board ship it is often very important to know the level of liquid in a tank.

(*a*) PURGE SYSTEMS

These use a small flow of gas, usually air, which bubbles slowly through the system, as indicated in Figure 3.5. Applying principles learned in physics:

Pressure indicated		Pressure due to height of fluid above the end of the tube		Pressure due to the head of gas in the vertical pipe		Pressure drop due to flow of gas between A and B
	=		−		+	

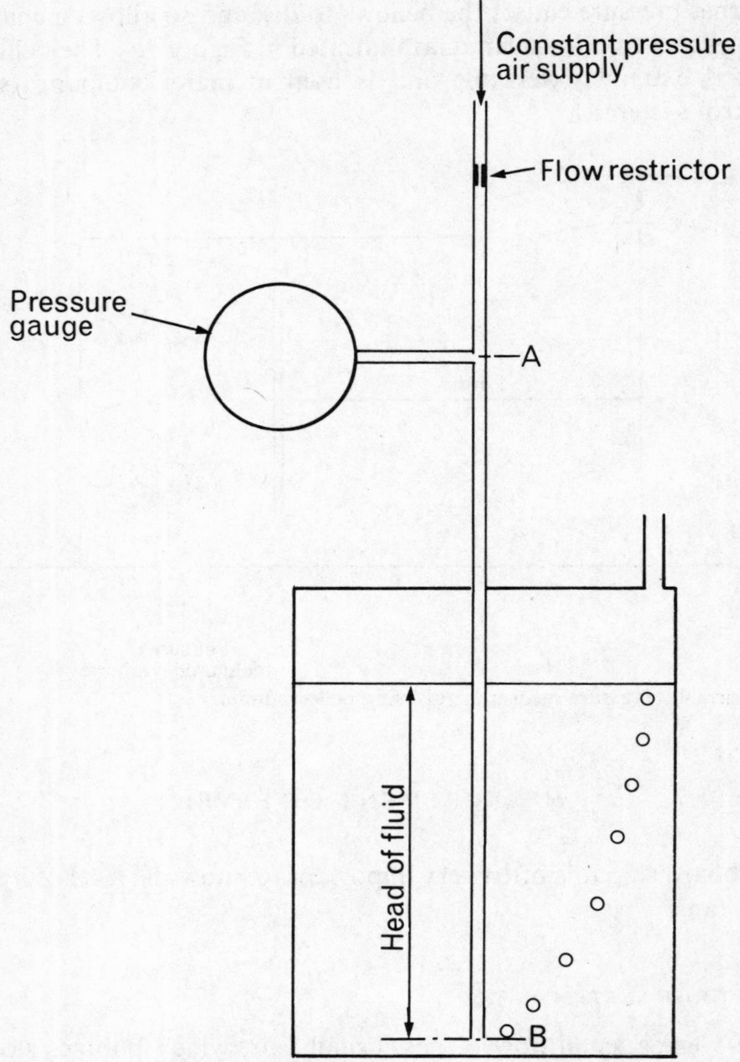

Figure 3.5 Measurement of level by a purge system.

Neglecting the last two terms:

Pressure indicated	= Specific weight of fluid in the tank	×	Height of fluid above the open end of the purge pipe
Units: [N/m²	= N/m³	×	m]

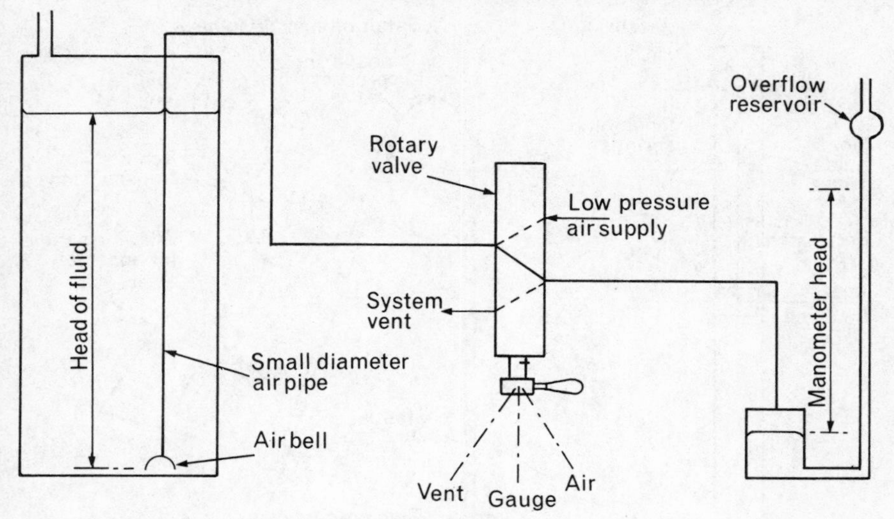

Figure 3.6 Remote tank-level measuring system.

A commonly fitted system using this general principle is shown in Figure 3.6 of the remote tank level-measuring system. Enough air is blown into the purge pipe to completely clear the pipe and air bell of the tank fluid, and the purge pipe is then connected to the manometer. The pressure exerted by the fluid in the tank compresses the air in the bell, and this pressure is communicated to the manometer. In the equilibrium position:

Height of fluid in the tank above the bell	×	Specific weight of fluid in the tank	=	Difference in manometer levels	×	Specific weight of fluid in the manometer

If mercury of specific gravity 13·6 were used in the manometer, and if oil of specific gravity 0·9 were in the tank, a reading of 1·0 m on the manometer would represent 15·1 (= 13·6 ÷ 0·9) m in the tank, provided that the temperatures of the fluids were similar.

(*b*) FLOAT-OPERATED SYSTEMS

One method is shown in Figure 3.7. The float is attached to a graduated and perforated stainless steel tape. The tape is guided over a measuring sprocket to a tape drum, which maintains a constant tension in the tape. The measuring sprocket drives a counting mechanism, which can be read locally. In addition, it is

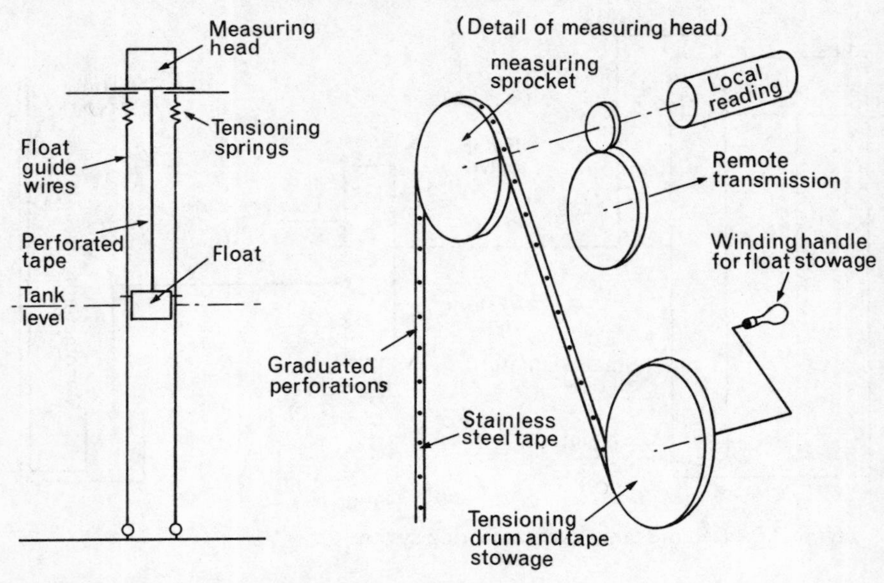

Figure 3.7 Float-operated level-measuring system.

usual for this information to be electrically transmitted to a cargo control centre.

(*c*) FLOAT-OPERATED LEVEL SWITCHES

These are often used on board to initiate alarms, to start and stop pumps, or to open and shut valves. A type illustrated by Figure 3.8 may be used. The float assembly carries with it a permanent

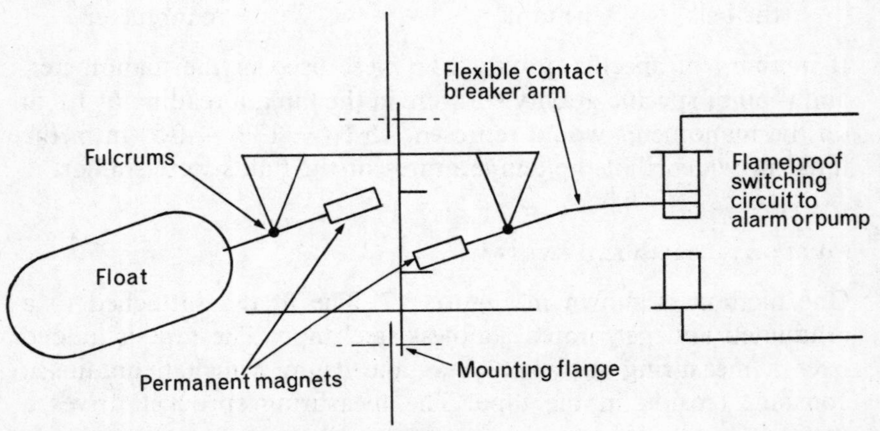

Figure 3.8 Float-operated level switch.

magnet, which is opposed by a similar magnet, operating an electrical or a pneumatic switching unit. The use of magnets which are arranged to repel each other gives the mechanism a snap action. It will be noted that the switching mechanism is completely separated from the space being protected.

MEASUREMENT OF FLOW

(*a*) VENTURIMETER

This device uses the principle of energy conservation. Referring to Figure 3.9:

$$\text{Total potential energy at 1} + \text{Kinetic energy at 1} = \text{Total potential energy at 2} + \text{Kinetic energy at 2}$$

neglecting the small losses between 1 and 2. From this equation it can be shown that:

$$\text{Mass flow rate} = \text{Density of fluid} \times \text{Constant for the meter} \times \sqrt{H}$$

This value of H can be measured, e.g. by a bellows unit, and the measured value transmitted to some remote position.

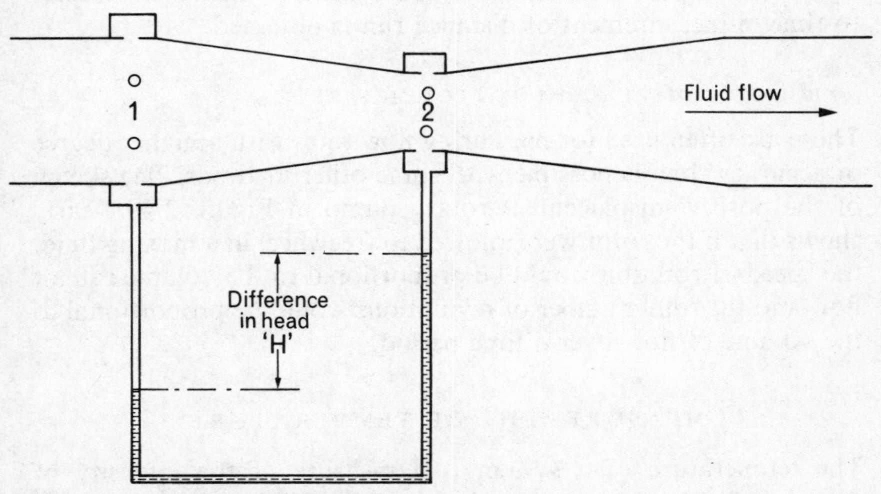

Figure 3.9 Principle of the venturimeter.

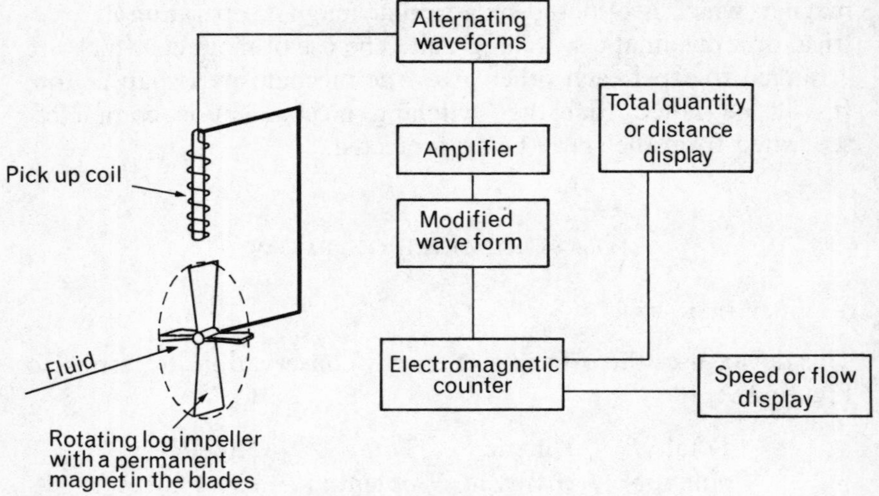

Figure 3.10 Turbine flowmeter or ship's log.

(*b*) TURBINE-TYPE FLOWMETERS

These use a rotor mounted in the tube carrying the fluid, so that its rate of rotation is proportional to the speed of flow. The rotor speed is sensed by the pick-up coil, as is indicated in Figure 3.10, and is processed in the electronic system.

This principle can be used to measure a ship's speed if the rotor is mounted in the water stream flowing beneath the hull of a ship. By integrating the instantaneous speed measurement with respect to time, a measurement of distance run is obtained.

(*c*) POSITIVE DISPLACEMENT FLOWMETERS

These are often used for measuring flow rates with a higher degree of accuracy than is possible with some other methods. The sketch of the positive-displacement rotary pump in Figure 2.40 clearly shows that if the rotor were allowed to freewheel in a moving fluid, the speed of rotation would be proportional to the volume rate of flow and the total number of revolutions would be proportional to the volume of flow over a time period.

MEASUREMENT OF TEMPERATURE

The temperature of a system is a measure of the intensity of activity of the molecules in that system. The temperature of a system is also measure of the heat energy which may be transferred to or from that system when it is in contact with surroundings of a different temperature. Hence 'heat' is a transient form of energy which crosses system boundaries, due to temperature differences. Figure 3.11 illustrates the difference between heat and temperature.

Surroundings at temperature T_{high}

System of:
MASS – m
SPECIFIC HEAT CAPACITY – C
TEMPERATURE – T_{low}

Boundary of system

Heat energy transfer to system equals
mass
x
specific heat capacity
x
rise in temperature of the system

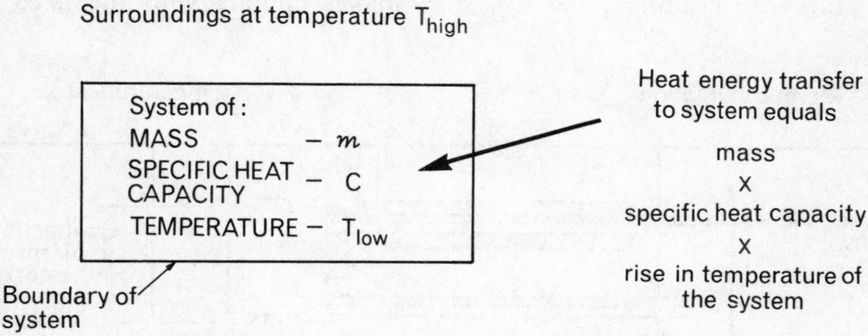

Figure 3.11 Difference and relationship between heat and temperature.

Temperature measurement is achieved by three main methods on board ship: expansion, electrical, and radiation methods.

Expansion Methods

(*a*) EXPANSION OF SOLIDS

When heat energy is transferred to a solid, the molecules of that solid absorb that energy and oscillate about their mean positions at a greater amplitude, which results in the expansion of that solid. This expansion has a predictable relationship with the temperature of the solid, involving its coefficient of expansion. This coefficient varies with different materials and forms the principle of operation of the bimetallic thermometer, which uses two materials wound together in the form of a closely-coiled helical spring. As the temperature of its surroundings changes, the materials expand at different rates, causing the spring to wind or unwind. This movement can operate a pointer over a scale, and so indicate temperature. Such units are often used to measure temperatures in the exhaust gas manifolds of diesel engines, or cargo temperatures, e.g. of bitumen.

The diagram of the rate of temperature rise detector (Figure 3.12) illustrates another example of the use of the bimetallic strip, in this case in a fire detection system. Should a fire start in the space protected by the device, heat energy is transferred to the strips. Should the rate of temperature rise be slow, there is enough time for heat energy to pass through the thermal insulation so that the strips will bend upwards together. Should the slow temperature rise continue, the top strip makes contact with the fixed

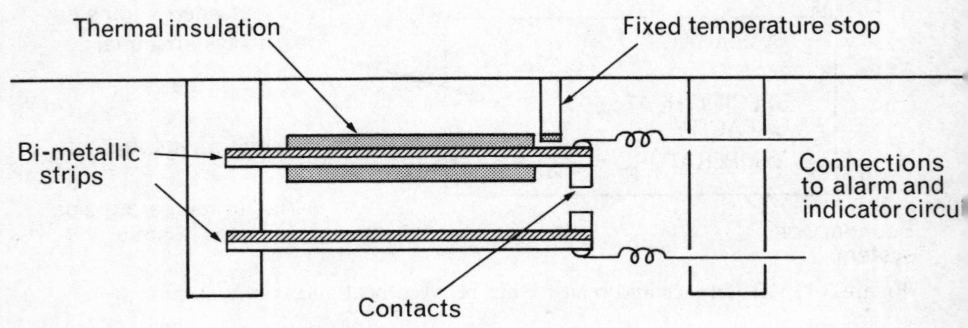

Figure 3.12 Rate of temperature-rise detector.

temperature stop, allowing the contacts to close and so instigating an alarm. Should the rise in temperature be rapid, the lower strip bends upwards much more quickly than the top strip, so instigating the alarm very quickly.

(*b*) EXPANSION OF LIQUIDS

In the well-known liquid-in-glass thermometer, a liquid, usually mercury, is contained within a glass bulb and extends into a small-diameter tube. Clearly, if heat energy is transferred to the thermometer, the mercury expands at a greater rate than the glass and rises up the tube, indicating temperature. The great disadvantage of this device is the difficulty experienced in transmitting the information to some remote position.

This difficulty is overcome by using a mercury-in-steel thermometer, as shown in Figure 3.13. A change in temperature causes heat energy to flow into the bulb where the mercury expands at a faster rate than the bulb. The expansion of the mercury causes the Bourdon tube to uncurl, moving the pointer over the

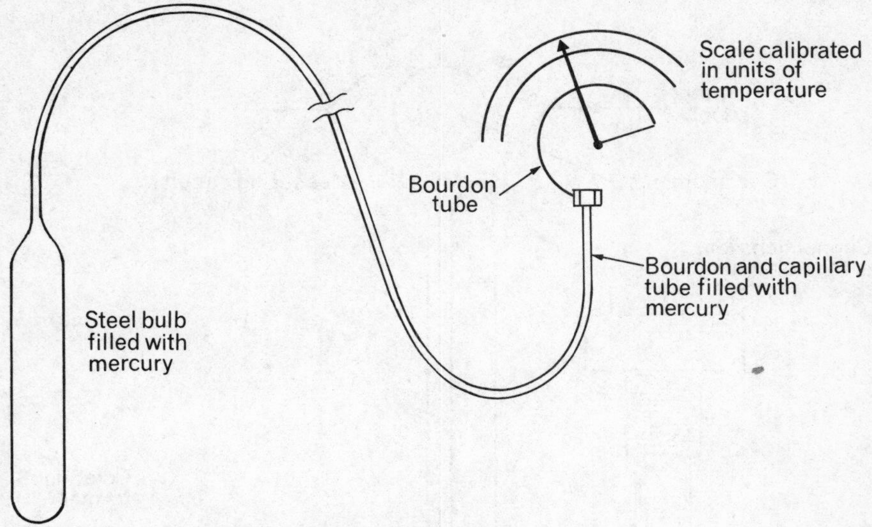

Figure 3.13 Mercury-in-steel thermometer.

graduated scale. There is a limit to the length of the tube, since the temperature of the tube surroundings affects the temperature reading.

(*c*) EXPANSION OF GASES

If a gas were contained in a system similar to that just described, instead of liquid, the result would be essentially a constant volume system. Hence temperature changes would be accompanied by pressure changes, according to the gas laws, which could be detected by a Bourdon tube. This is the principle of the gas thermometer. The gas-filled bulb is generally much larger than the liquid-filled bulb and its response is more sluggish.

Electrical Methods

(*a*) RESISTANCE THERMOMETER

From studies in physics it is known that the resistance of pure metallic conductors increases with temperature. If a resistance coil is used as a temperature sensor, as indicated in Figures 3.14 and 3.15, changes in temperature will be accompanied by predictable

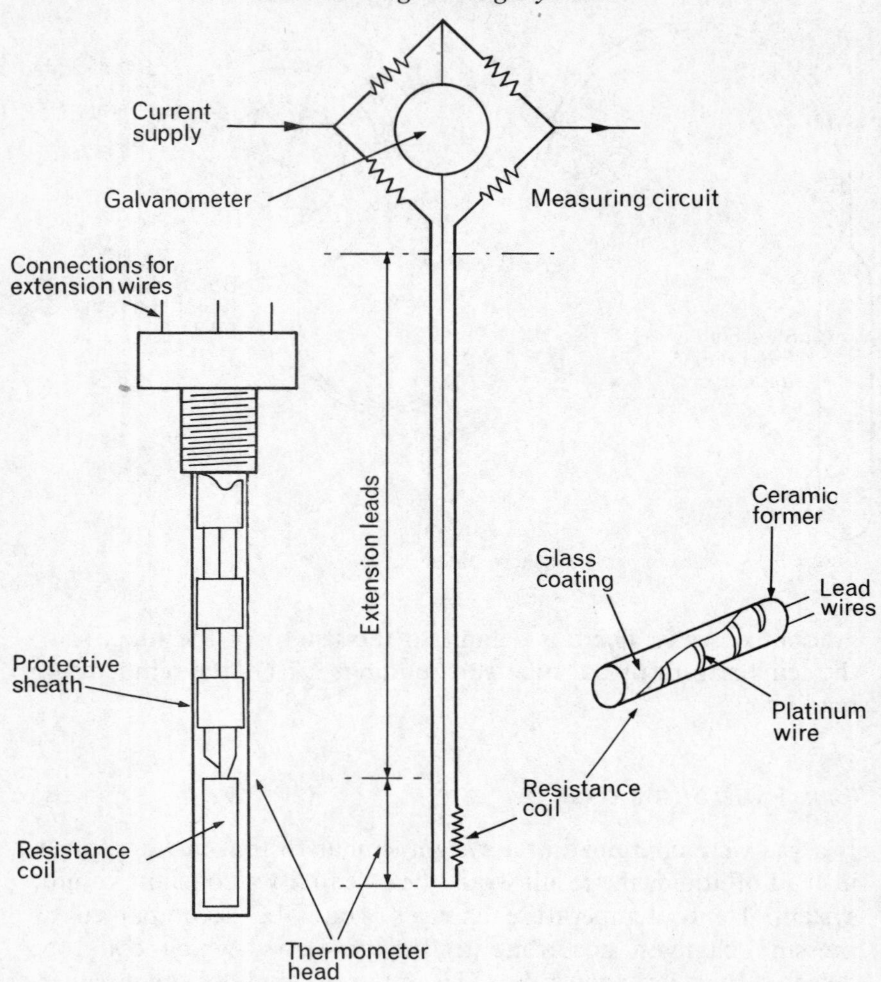

Figure 3.14 Resistance thermometer.

changes in the resistance of the coil, which can then be measured by the appropriate circuitry.

(*b*) THERMISTORS

Thermistors consist of a small bead of semiconductor materials which have a temperature coefficient which is about ten times greater than that of the copper or platinum resistance coil which they replace. They have a very small mass; hence they respond rapidly to a change in the temperature of the measured variable.

However, they may be subject to drift over a period of time and they are subject to wider manufacturing tolerances than are other devices. The thermistor temperature–resistance characteristic is shown in Figure 3.15.

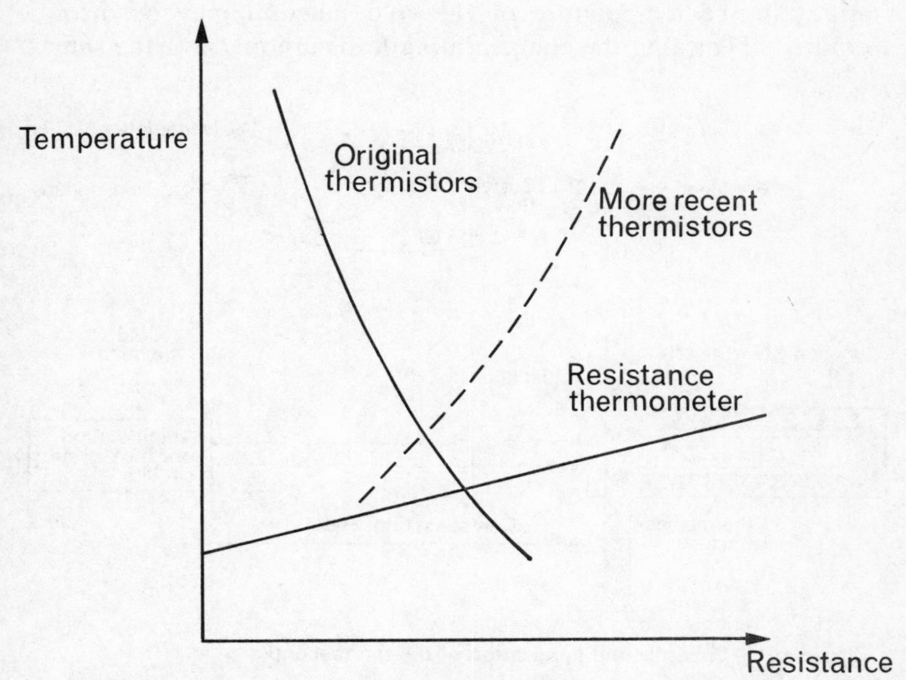

Figure 3.15 Temperature–resistance characteristics of the resistance thermometer and thermistor.

(*c*) THERMOCOUPLES

If a circuit is constructed of two different metals, and one junction is maintained at a higher temperature than the other, a current will flow around the circuit, as indicated in Figure 3.16, depending on the temperature *difference* between the junctions. The electromotive force causing this current flow is predictable. Hence this principle can be used to measure temperatures on board ship, as is also indicated in the diagram.

If the compensating leads were made of the same material, the effective cold junction would be at the junction box and the instrument would indicate the difference between the measured variable temperature and the temperature at the junction box, which would be small. For effective measurement, compensating leads, with similar thermo-electric properties to those used in the thermocouple, are used to extend the cold junction to some position at a more constant temperature, probably in a centralized control station. Additionally, circuitry to compensate for variations in the temperature of the cold junction may be incorporated. Typically, the compensating leads are made of the same

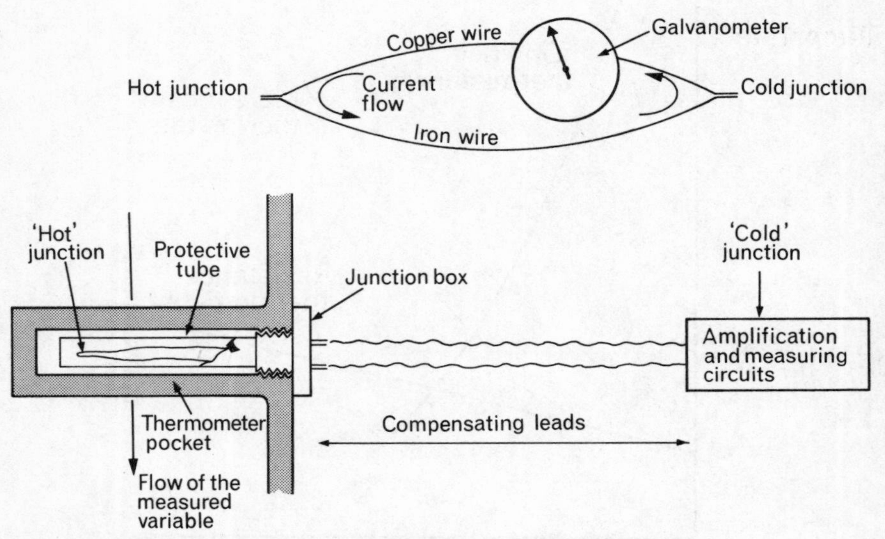

Figure 3.16 Principle and application of the thermocouple.

metals as are used in the measuring head, but they are manufactured to lower standards of purity, which reduces their cost considerably.

Radiation Thermometers

Energy is radiated from a hot body at a rate which is dependent on the absolute temperature of that body. The distribution of the radiated energy within the spectrum is illustrated in Figure 3.17. In particular, it can be seen that both the frequency and emission rate of the radiated energy depend on the surface temperature of the body. In order to measure radiation of different frequencies and intensities, different types of photo-electric cell may be used.

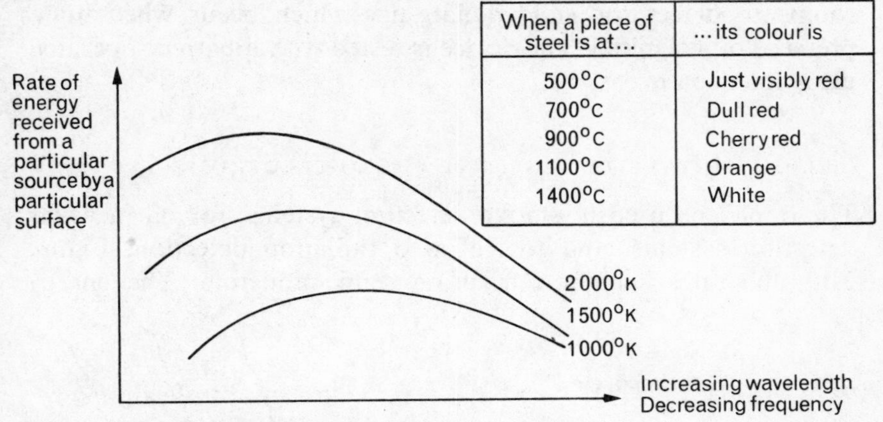

When a piece of steel is at...	... its colour is
500°C	Just visibly red
700°C	Dull red
900°C	Cherry red
1100°C	Orange
1400°C	White

Figure 3.17 Temperature–radiated energy spectrum.

(*a*) FREQUENCY-SENSITIVE PHOTO-ELECTRIC CELL

Such a cell is shown in Figure 3.18. When radiated energy falls on to the specially coated cathode, electrons absorb the energy, which may be enough to allow some of them to escape from the surface of the cathode. The effectiveness of this process increases with the frequency of radiation, and there is a sharply-defined critical lower limit of frequency for each particular coating material.

Hence a device can be constructed which is sensitive to ultraviolet light but not to the lower frequency of visible light. A small potential applied to the anode attracts the free electrons, producing a current flow when the frequency of the radiated energy exceeds a certain value. These devices are used in engine

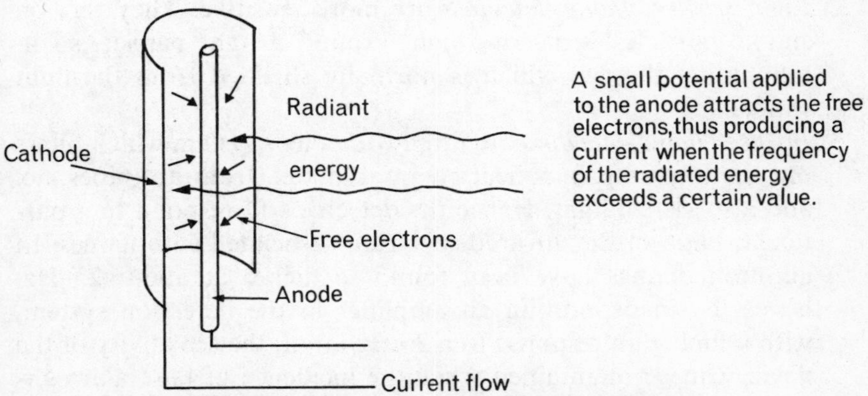

Figure 3.18 Principle of a frequency-sensitive photo-electric cell.

rooms to detect the sudden flare-ups which occur when high-pressure oil is ignited. The device is tested with a battery-operated ultraviolet torch.

(*b*) LIGHT-INTENSITY-SENSITIVE PHOTO-ELECTRIC CELLS

These may be used in smoke detection systems, for oil in water detection systems, and for infrared radiation detection. Figure 3.19 illustrates such a cell, using semiconductors. The energy

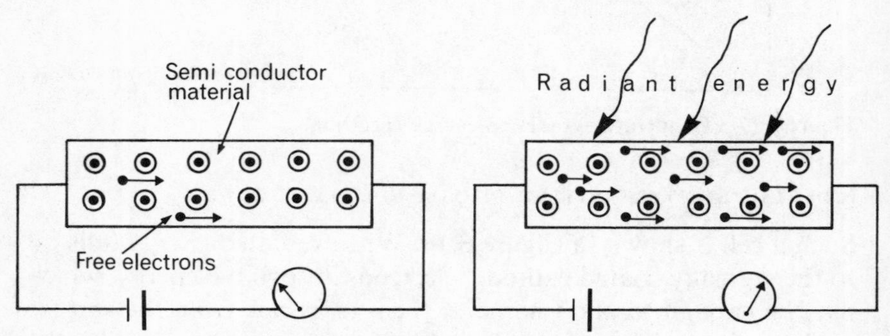

Figure 3.19 Photo-electric cell sensitive to light intensity.

radiated increases the number of mobile electrons, and under suitable conditions the current flow is proportional to the intensity of the radiation. This principle can be utilized in different ways:

1. *Observation-type smoke detectors* rely on smoke's obscuring the light emitted from a lamp, so reducing the output from the photocell and causing an alarm to sound.
2. *Light-scatter smoke detectors* are more sensitive. They rely on smoke particles' scattering light around a light barrier, so illuminating the cell which is normally shielded from the light source.
3. *Infrared flame detectors* are fitted with a lens system which filters out the low-frequency heat energy radiated from hot pipes etc. and also visible light. Hence the detector will respond to a particular band of the infrared spectrum associated with flames. In addition, flames have been found to flicker at about 25 Hz; hence, by incorporating an amplifier in the detection system, with a high gain response to a 25 Hz input, the sensitivity of the device can be maintained while the incidence of false alarms is reduced.

MEASUREMENT AND CONTROL OF HUMIDITY

SLING HYGROMETER

The sling hygrometer is shown in Figure 3.20. As the hygrometer is rotated, air flows past the thermometer bulbs. As moisture evaporates from the damp sleeve surrounding the wet bulb, it

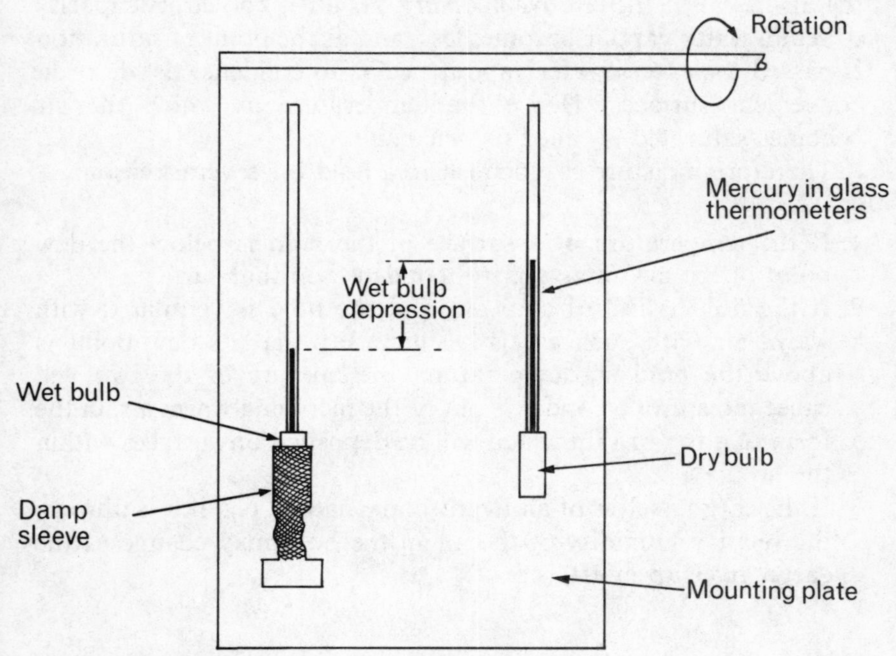

Figure 3.20 Sling hygrometer.

cools the wet thermometer, which shows a reading below that of the dry thermometer. The rate of moisture evaporation depends on the capacity of the air to hold more moisture; hence the difference between the readings is a measure of the humidity of the air in which the hygrometer is swung.

APPLICATION TO THE CONTROL OF HUMIDITY IN A CARGO HOLD

The evaporation and condensation of water in a cargo hold may cause rusting, staining, corrosion, or mildewing of the cargo and hold. To prevent this it is necessary to measure and control the moisture content of the air in the hold.

The amount of water vapour which can be absorbed by air increases with the air temperature, but there is always a limit, and when this is reached the air is said to be *saturated.* At any point below saturation, the proportion of water vapour actually contained in the air, compared with the amount required to saturate the air, is called the *relative humidity*. As air is cooled its capacity to retain water vapour becomes less, and as the point of saturation is passed the excess water vapour begins to condense as 'dew' on convenient surfaces. Hence the temperature at which the air becomes saturated is called its *dew point*.

Therefore moisture can deposit in a hold for several reasons:

1. If the temperature of a surface in the hold is below the dew point of the air, moisture will condense on that surface.
2. If the hold is full of cool air, and the hold is ventilated with warm air with such a relative humidity that its dew point is above the hold air temperature, the mixture of the two will cause moisture to condense out of the incoming warm air, in the form of a fog or rain which will be deposited on surfaces within the hold.
3. If the cargo itself is of an hygroscopic nature, e.g. if it is fibrous, the relative humidity of the air in the hold may change as the cargo gives up moisture.

PREVENTION SYSTEMS

A typical prevention system may have as its object the control of the dew point of the hold air. From the notes above it can be seen that this can be accomplished by:

1. Controlling the relative humidity of the air in the hold.
2. Controlling the temperature of the air in the hold.

It is expensive to control the temperature; therefore method 1 will be described, which has as its aim *the control of the dew point of the air in the hold so that it is below the temperature of the hull and cargo*, which are assumed to be at or near the sea water temperature.

A humidity control system is illustrated in Figure 3.21. One drying unit may be provided for a group of cargo spaces, and air from the hold can be circulated over the drying unit, which lowers its relative humidity and therefore its dew point at a given pressure and temperature. Outside air can also be introduced into the hold by the fan if ventilation is required. The drying material in the rotating heat exchanger may be a heat-resistant fibrous material,

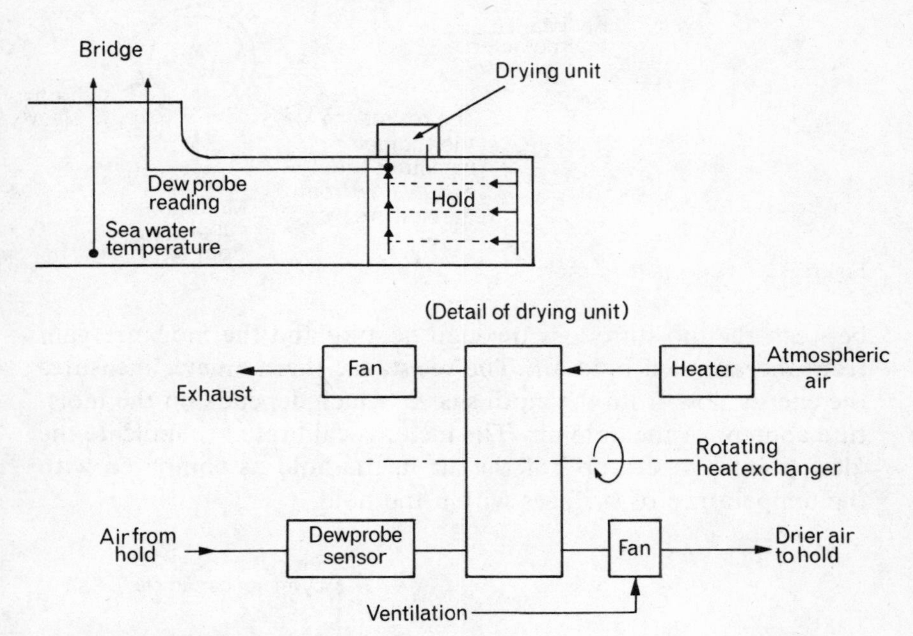

Figure 3.21 Humidity control system.

impregnated with a hygroscopic chemical. The moisture extracted from the hold air is removed from the drying material by a heat supply, which may be electrical energy or steam.

The dew point (or dewprobe) sensor consists of a pair of gold alloy wires wound around a cloth tube, impregnated with a salt such as lithium chloride (*see* Figure 3.22). As the moisture content of the hold air passing over the dew point sensor increases, the resistance of the conducting path between the wires is reduced. The resulting increase in current flow heats up the salt, drying it, and so tends to reduce the heat flow from the dew point sensor to the temperature sensor, which is a resistance thermometer encased in a stainless steel tube. A position of equilibrium is reached

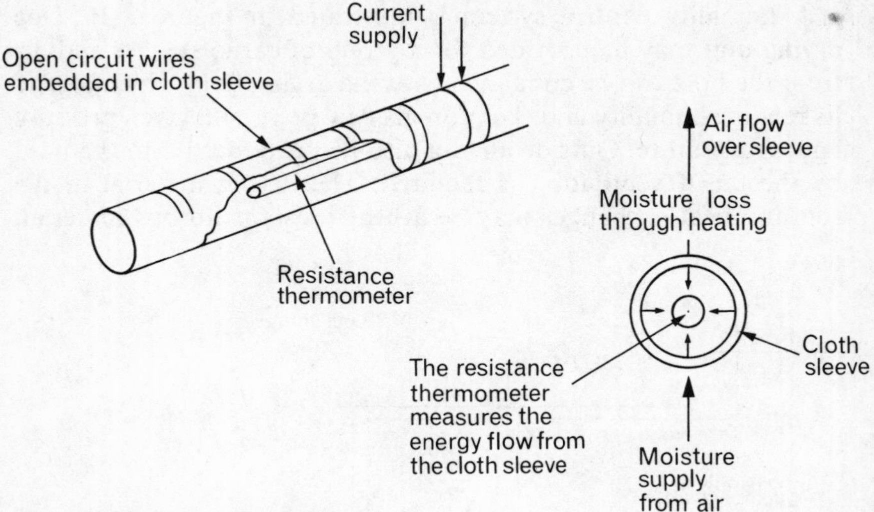

Figure 3.22 Dew point sensor.

between the moisture loss through heating and the moisture gain from the sampled hold air. The resistance thermometer measures the energy flow from the cloth sleeve, which depends on the moisture content of the hold air. The meter is calibrated to indicate the 'dew point temperature' of the air in the hold as compared with the temperature of surfaces within the hold.

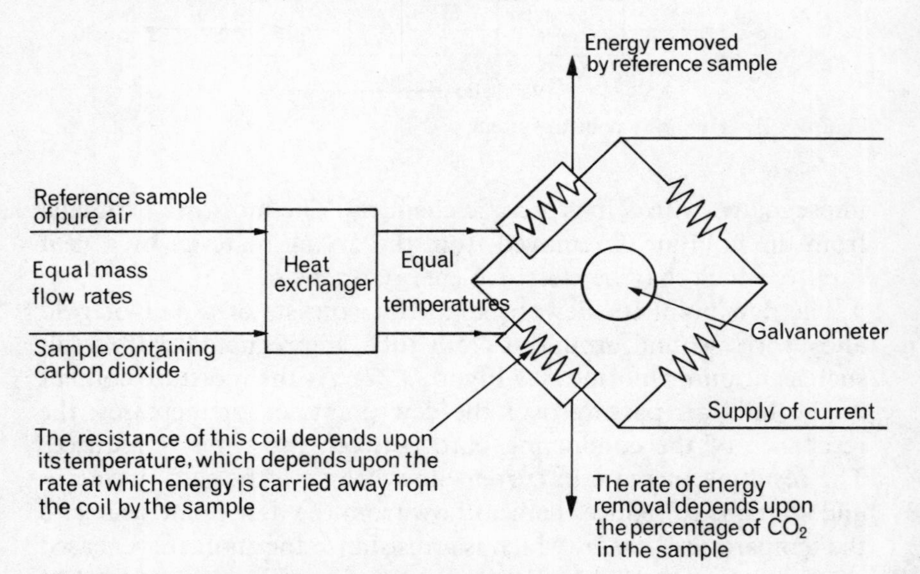

Figure 3.23 Measurement of carbon dioxide content.

MEASUREMENT OF CARBON DIOXIDE CONTENT

The measurement of the carbon dioxide content of a sample of air depends on the fact that the thermal conductivity of carbon dioxide is different from that of air.

Figure 3.23 illustrates a carbon dioxide detecting device. If the inlet temperature and mass flow of the measured sample and of the reference sample of pure air are similar, the temperature of the coils, and hence their resistance, increases until the rate of energy supplied by the current flow equals the energy carried away by the air and carbon dioxide flowing past each part of the Wheatstone bridge. This depends on the relative composition of the samples, and therefore the current flowing through each arm of the bridge will unbalance, allowing the galvanometer to be calibrated in terms of percentage of carbon dioxide in the measured sample.

Such measurement systems are used to determine the carbon dioxide content of flue gases and inert gas systems, and more sensitive types are used to measure the carbon dioxide in refrigerated fruit cargo spaces.

MEASUREMENT OF HYDROCARBON GAS CONTENT

The detection and measurement of the percentage of hydrocarbons in air is important, in order to maintain the safe operation of the ship. An appropriate instrument is shown in Figure 3.24. A

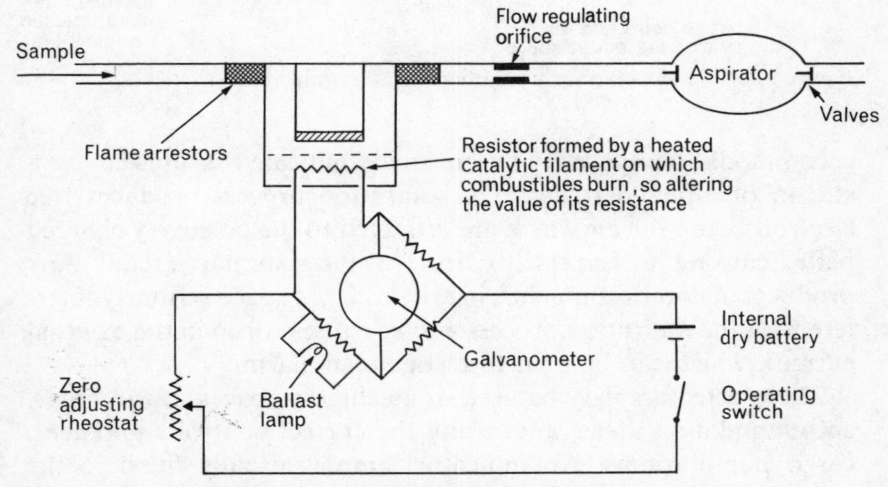

Figure 3.24 Measurement of hydrocarbon gas content.

sample is drawn into the instrument, at a constant volume flow rate, by the aspirator. As the sample passes over the catalyst, heat is produced by a chemical reaction, which alters the resistance of the coil. This alters the balance of the Wheatstone bridge measuring circuit, the instrument usually indicating the amount of hydrocarbon gas as a percentage of the lower explosive limit.

DETECTION OF COMBUSTION PRODUCTS

In the early stages of a fire, minute particles become airborne before visible quantities of smoke are discernible. The device illustrated in Figure 3.25 has been designed to detect these and to instigate an alarm in a fire detection system in engine rooms or

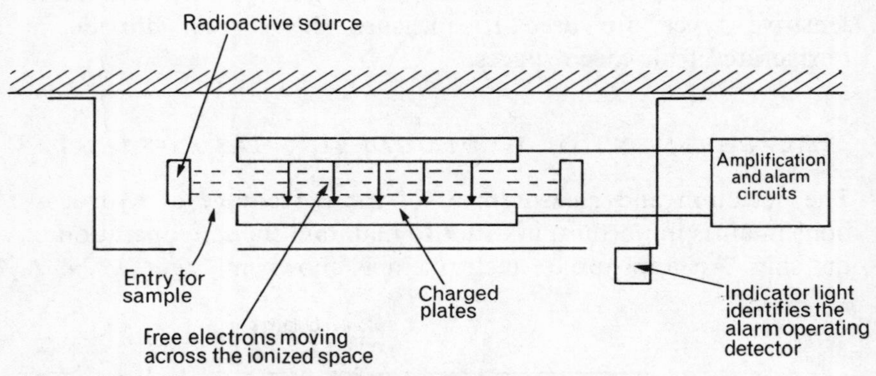

Figure 3.25 Smoke detector using the ionization principle.

accommodation spaces. The air space indicated is ionized by a stream of alpha particles. The ionization process produces free electrons from the air which are attracted to the positively charged plate, causing a current to flow in the external circuit. Any products of combustion entering the charged space seriously interfere with the ionization process and so cause a drop in the external current, which can be used to instigate an alarm.

These detectors may be used in machinery spaces, cargo holds, accommodation areas, and, using the correct control equipment, cargo pump rooms. An indicator lamp is usually fitted in the device, to indicate the unit which has instigated an alarm.

MEASUREMENT OF OXYGEN CONTENT

The measurement of the oxygen content of a gas sample is important for the safe operation of inert gas systems.

Oxygen has the rare property that it is very susceptible to magnetic attraction, compared to other gases such as nitrogen and carbon dioxide, provided that the oxygen remains cool. This fact is used in the measuring device shown in Figure 3.26.

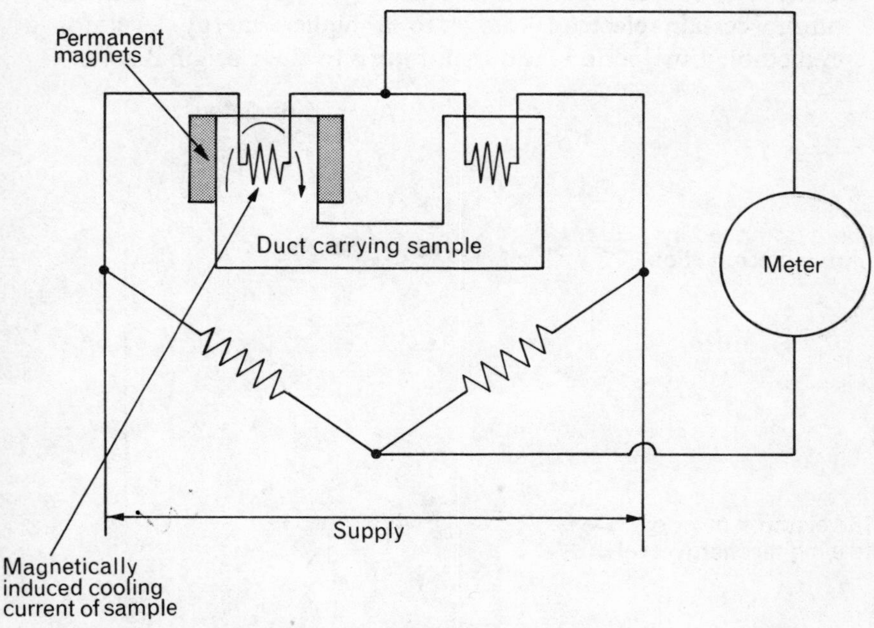

Figure 3.26 Measurement of oxygen content.

The permanent magnet attracts the oxygen around the left-hand resistance coil. As the oxygen cools the resistor, it itself becomes warmer and loses some of its magnetic susceptibility, and is forced away from the resistor by cooler oxygen which is still attracted by the magnetic field. In fact, a magnetically induced draught is established, which cools off the left-hand resistor at a rate dependent on the percentage of oxygen in the sample. The variation in cooling rate, and hence the oxygen content, is measured by an electrical measuring circuit, such as a Wheatstone bridge.

MONITORING OIL IN WATER

The current IMCO regulations limit the oil discharge rate of a ship to 60 l. per mile, with an additional limit on total discharge per voyage. A continuous recording of the oil content of the discharged ballast provides the operator with proof that the statutory requirements have been met.

One method of measurement uses the physical phenomenon of *fluorescence*, which is illustrated in Figure 3.27. If the atoms of certain substances receive radiated energy from a high-frequency source, certain electrons move to a higher energy level for a predictable time period, and then return to their original state. On

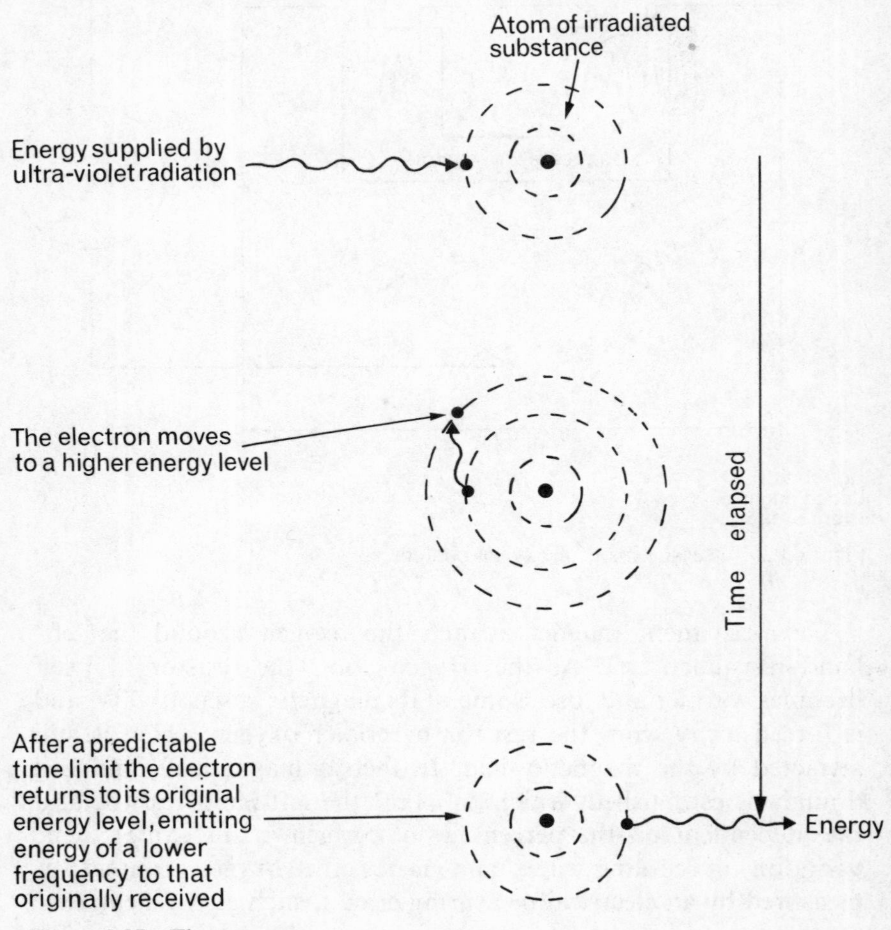

Figure 3.27 Fluorescence.

returning to a lower energy level, they emit energy at a frequency lower than that of the energy originally received.

Oil fluoresces more easily than water, a fact which provides the means of operation for the detection system shown in Figure 3.28.

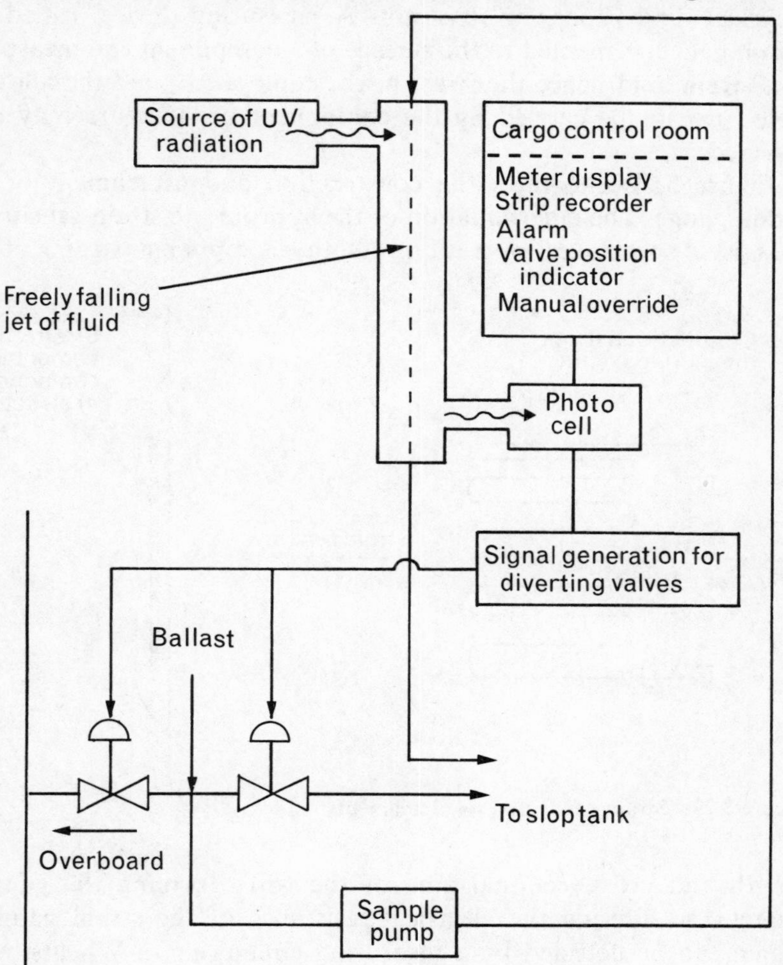

Figure 3.28 Oil-in-water monitoring system.

A sample from the overboard discharge receives energy from the ultraviolet lamp and falls freely through a tube. The energy emitted due to any fluorescence is monitored by a photo-electric cell, which produces a signal dependent on the presence of oil in the sample. In the event of excess contamination, the diverting valves, shown in the diagram, are automatically operated.

MEASUREMENT OF STRESS

When a component or structure is subjected to a force it is said to be in a state of stress, which causes the component to change its physical dimensions, or to strain. A measuring device, called a strain gauge, cemented to the surface of a component can measure this strain, and hence the stress in the component, and the allowable force to be carried by the component or structure may be inferred.

Figure 3.29 illustrates the construction and attachment of a strain gauge. The determination of the appropriate strain-sensitive axis is a specialized operation. As the component strains, the

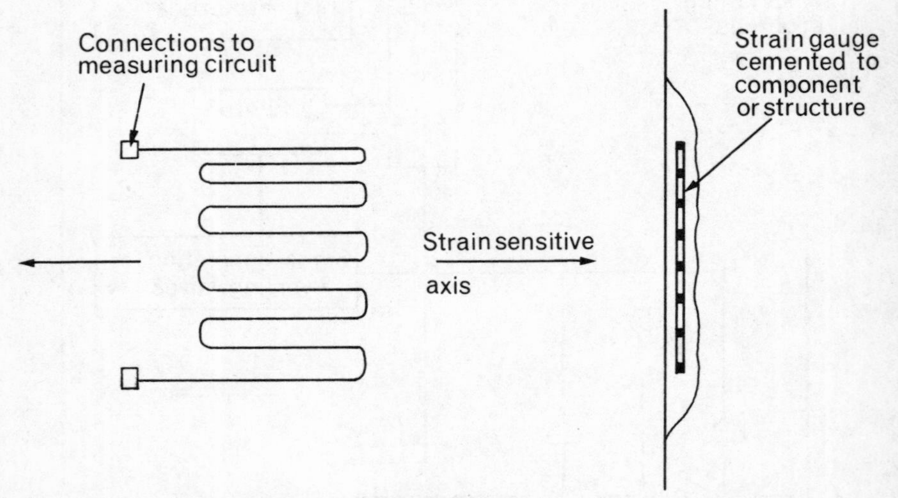

Figure 3.29 Strain gauge and its attachments.

length and cross-sectional area of the wires forming the gauge change, so altering the electrical resistance of the strain gauge, which can be detected by a measuring circuit, e.g. a Wheatstone bridge.

These gauges are now being fitted on ships to monitor the stresses in the ship structure which are due to cargo distribution and wave action. In certain very advanced ships, strain gauges are part of the automatic closed-loop system controlling tanker discharge and loading. They are also used to provide warning of overloads in deck cranes, to read cable tensions on cable layers, and in automated berthing systems.

MEASUREMENT OF TORQUE

When a shaft is subject to a torque, a twist develops in that shaft. This twist, or angular deflection, is proportional to the applied torque and to the torque transmitted. A pattern of strain gauges fixed to the rotating shaft can detect this 'angular strain', and hence the torque in the shaft can be assessed.

MEASUREMENT OF SHIP SPEED

Figure 3.30 illustrates how an electromagnetic principle can be used in measuring the speed of a fluid, relative to electrodes. The magnetic field is generated by a probe mounted beneath the ship.

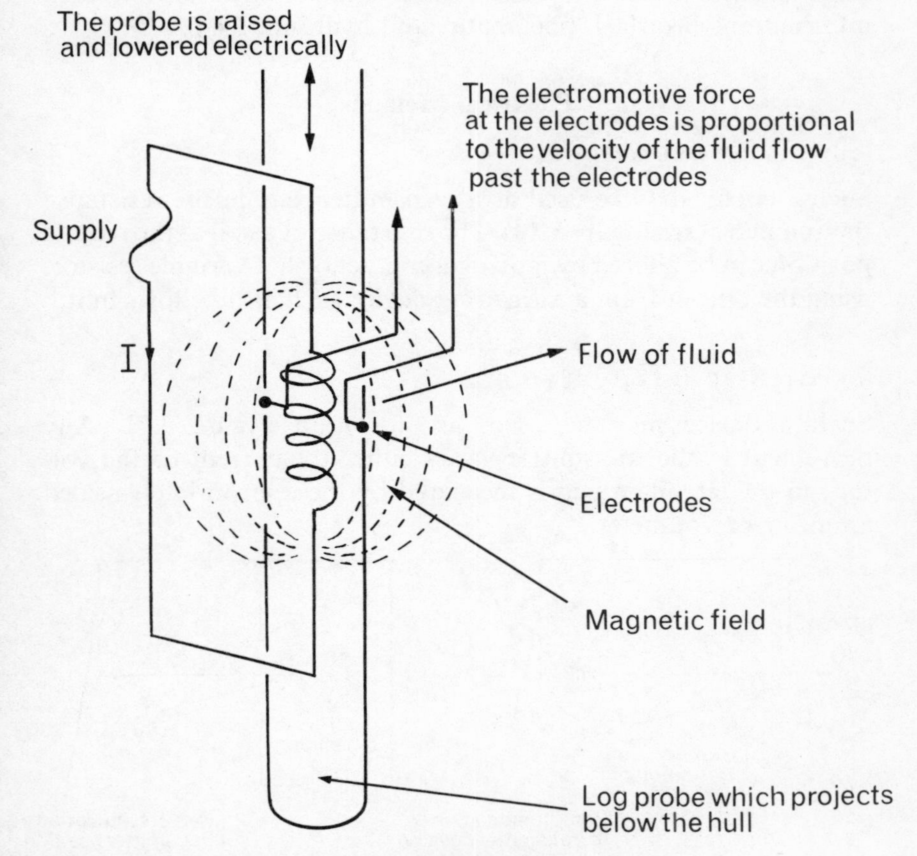

Figure 3.30 Electromagnetic measurement of ship speed.

The potential difference generated by the water flowing through the field is sensed by electrodes mounted on the surface of the probe. The output is processed electronically and integrated to give distance.

A speed and distance indicator is fitted on the bridge and a speed indicator in the engine room. In addition, the processor generates a ship speed input for the radar.

A flowmeter may be constructed using the same principle.

REMOTE TRANSMISSION SYSTEMS

On board ship it is very often necessary to measure a variable in one position, and to transmit the information to another position some distance away. There are three methods of transmitting such information: electrical, pneumatic, and hydraulic.

Electrical Methods

(*a*) WHEATSTONE BRIDGE

Such a circuit may be used as a transmitter, e.g. in the resistance thermometer (*see* Figure 3.14). The resistance of a coil at the remote position can be altered by moving a contact over a variable resistor, using the output from a measuring device such as a bellows unit.

(*b*) POTENTIOMETRIC DEVICE

Such a device may be used, as shown in Figure 3.31. Any movement in the transmitter varies either the current or the voltage in the circuit, which is measured by the appropriately scaled ammeter or voltmeter.

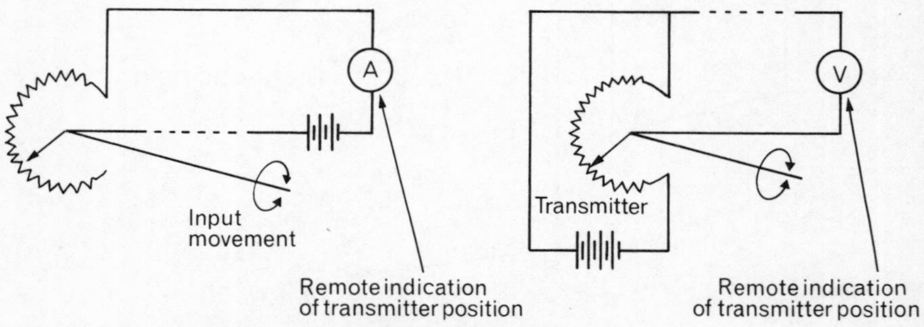

Figure 3.31 Potentiometric method of transmission.

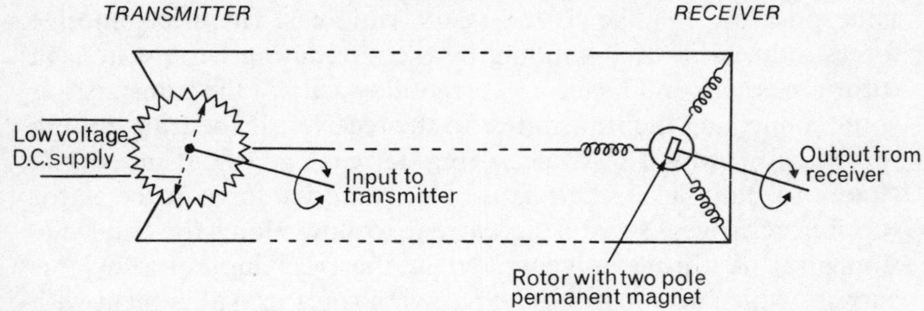

Figure 3.32 Transmission by direct-current position motor.

(*c*) DIRECT-CURRENT POSITION MOTOR

Transmission by this type of motor is illustrated in Figure 3.32. The transmitter resistor is tapped at three equidistant points. Sliding contacts, operated by the input to the unit, feed current into the input resistor, so establishing a certain current distribution, and hence a unique magnetic field in the region of the receiver windings. A change in the position of the current-distributing contacts in the transmitter therefore alters the magnetic field in the receiver, which is sensed by a two-pole permanent magnet mounted within the receiver windings, which rotates to a position determined by the position of the transmitter contacts. This arrangement is usually called a Desynn transmission system.

(*d*) ALTERNATING-CURRENT POSITION MOTOR

Transmission by such a motor is indicated in Figure 3.33. Both rotors are supplied by the same a.c. source. If the rotors are in the

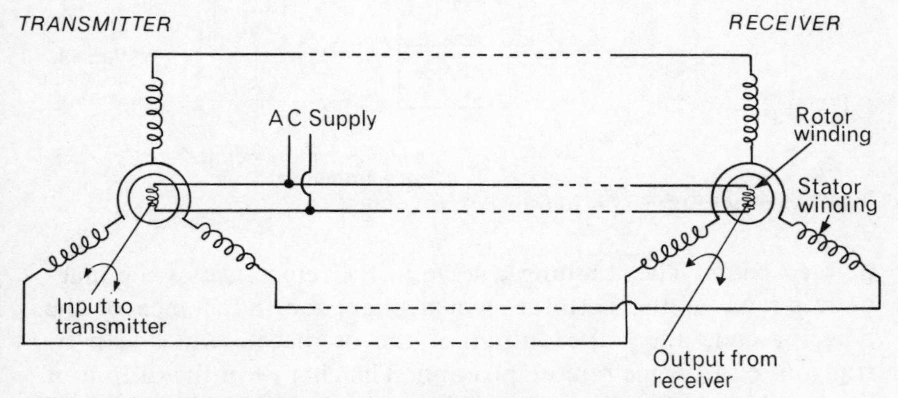

Figure 3.33 Transmission by alternating-current position motor.

same position, relative to the stator windings, the electromotive forces induced in each winding by the alternating current in each rotor are equal, and hence no current flows along the transmission wires connecting the transmitter to the receiver. If the transmitting rotor is now turned by a measuring device, e.g. a float measuring system for tank level, the relative electromotive force in the stator windings changes, so causing current to flow along the transmission lines. A torque is generated in the receiving rotor by this current, which acts to bring the receiving rotor into alignment with the transmitting rotor. This arrangement is called a Synchro, or Magslip, transmission system.

(*e*) FORCE–CURRENT TRANSMITTER

This is shown in Figure 3.34. The input force, in this instance generated by a pressure-measuring Bourdon tube, deflects the

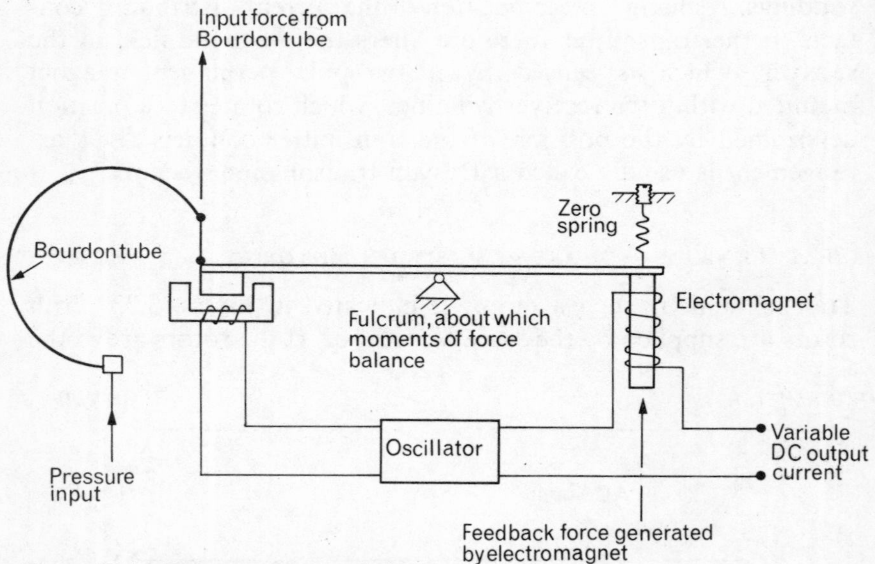

Figure 3.34 Force–current transmitter.

pivoted beam. The resulting change in the reluctance of the magnetic circuit of the detector, which alters the inductance of the detector coil, alters the output of the oscillator, which can be transmitted to some remote position. The change in the output of the oscillator is led through an electromagnet, which generates a negative feedback force, opposing the input force, so ensuring the stability of the transmitter.

Pneumatic Methods

Pneumatic methods often use a flapper–nozzle device to detect changes in measured variables, as indicated in Figure 3.35.

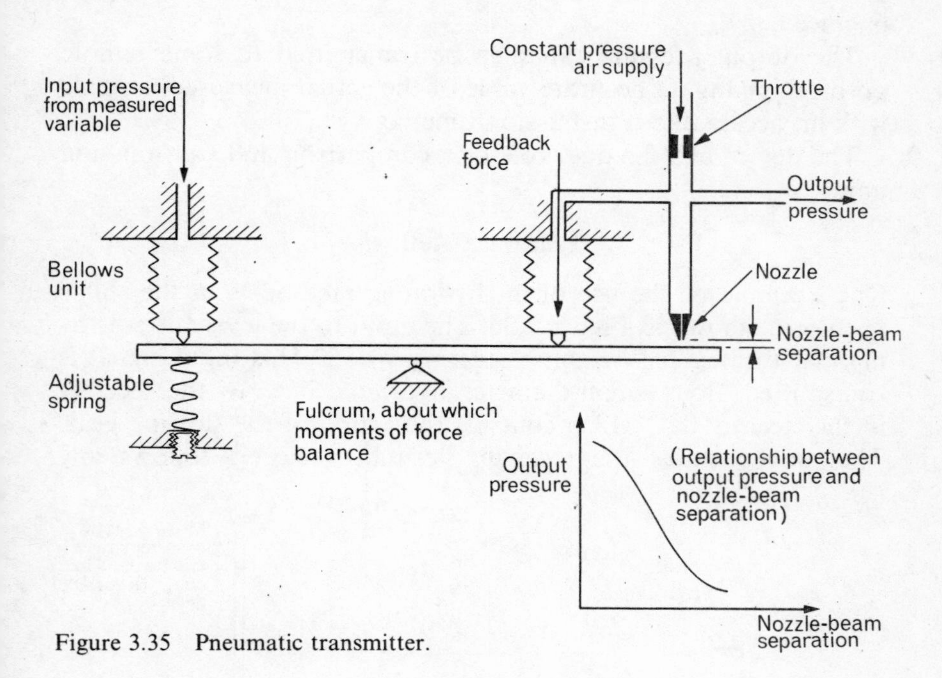

Figure 3.35 Pneumatic transmitter.

In this system a measurement of a controlled variable is obtained by means of a pressure signal, which could, for example, be derived from a tank-level measuring system using the purge principle. If the measured variable is at its desired value, the force it generates in the input bellows is exactly balanced by the force generated in the feedback bellows, and by the force in the spring, which depends on the set or desired value. For this condition there is a certain position for the beam, which means that the nozzle is separated from the beam by a definite predictable distance, which allows a predictable back pressure to build up downstream of the throttle in the supply line.

Since the relationship between nozzle separation and back

pressure is linear, over the operating range, the device transmits an output pressure which varies in proportion to changes in the measured variable, over the operating range. If the value of the measured variable increases, e.g. because of an increase in tank level, the left-hand end of the beam moves down and the beam moves closer to the nozzle, so increasing the back pressure. This back pressure acts on the feedback bellows, opposed by the spring, so that the beam assumes a predictable position and all forces are balanced.

The output pressure can then be transmitted to some remote position, giving an accurate value of the actual measured variable with an acceptable transmission time lag.

The device fills the dual role of a comparison and transmission unit.

Hydraulic Methods

One example of the use of an hydraulic method is in the ship's telemotor, shown in Figure 3.36. The input to the transmitter is by manual turning of the ship's steering wheel. This input signal is transmitted along small-diameter hydraulic lines to the receiver in the steering flat, which controls the action of the steering gear. The movement of the receiving telemotor closely follows any

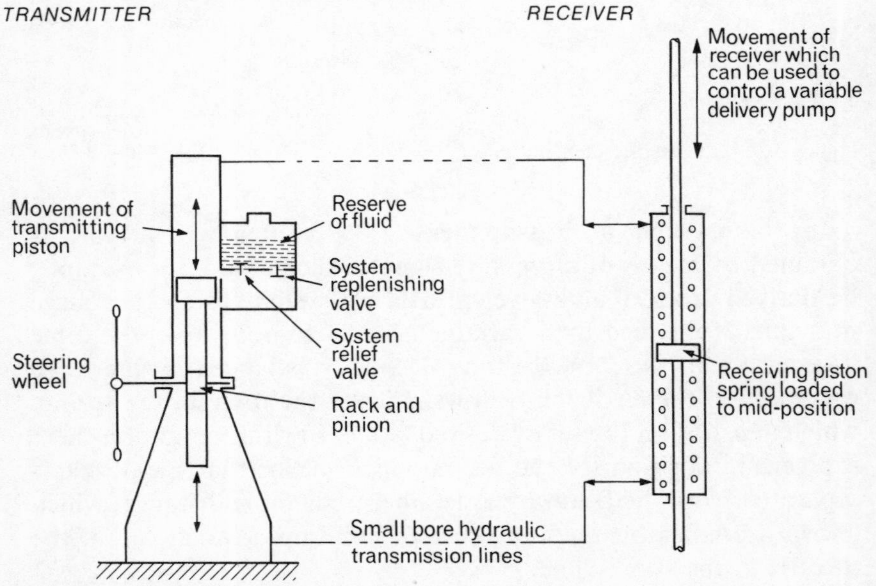

Figure 3.36 Ship's telemotor system.

movement of the transmitter, provided that there is no air and no leak in the system.

RESPONSE OF CONTROL SYSTEMS

Controllers receive the error signal from the transmission and comparison units, and manipulate this signal to achieve the required system response.

Suppose an ordinary glass thermometer is plunged quickly into a cup of hot water. The thermometer receives a 'step' input, but the output from the thermometer, i.e. its indicated reading, is different from the input, as is shown in Figure 3.37. The time lag is

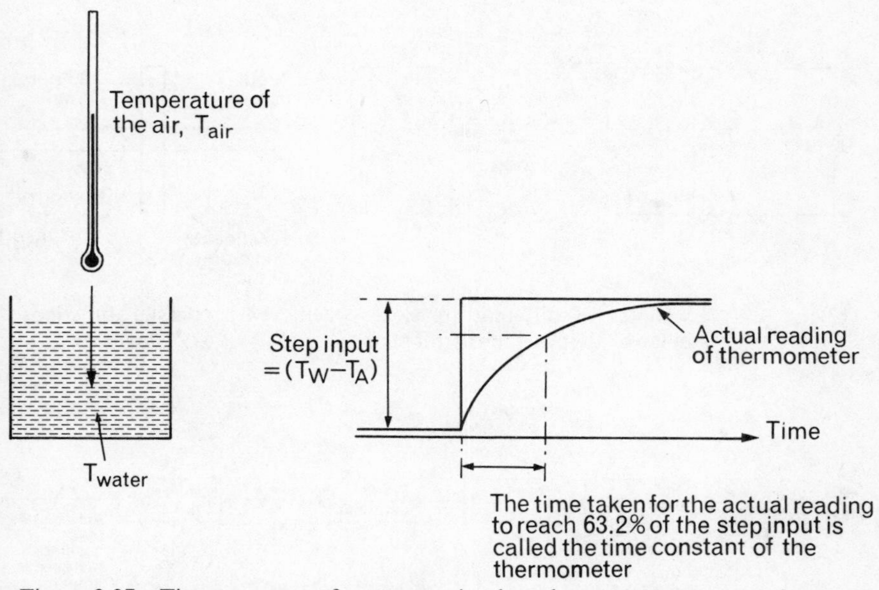

Figure 3.37 Time response of a mercury-in-glass thermometer to a step input.

due to the glass's resistance to heat transfer, and to the thermal capacitance of the mercury and glass. Hence a thick walled thermometer, with a comparatively large amount of mercury, has a longer time constant than a less massive thermometer.

If the thermometer is now subjected to a cyclic temperature input, its response will be as is indicated in Figure 3.38. Note that the output value, i.e. the value of the variable *as measured by the thermometer*, is less than the input value. It is attenuated and lags

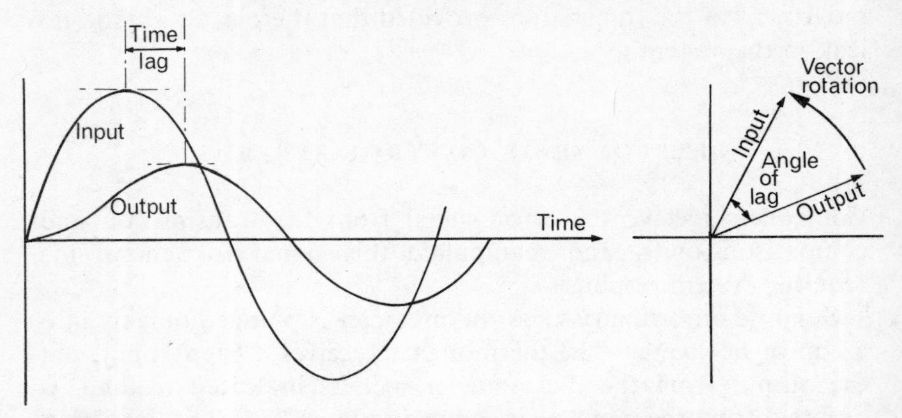

Figure 3.38 Time response of a mercury-in-glass thermometer to a cyclic input.

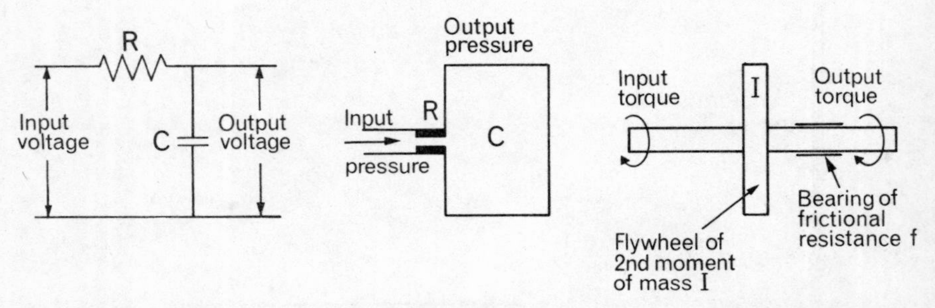

Figure 3.39 Examples of different physical systems with transfer functions similar to that of the thermometer.

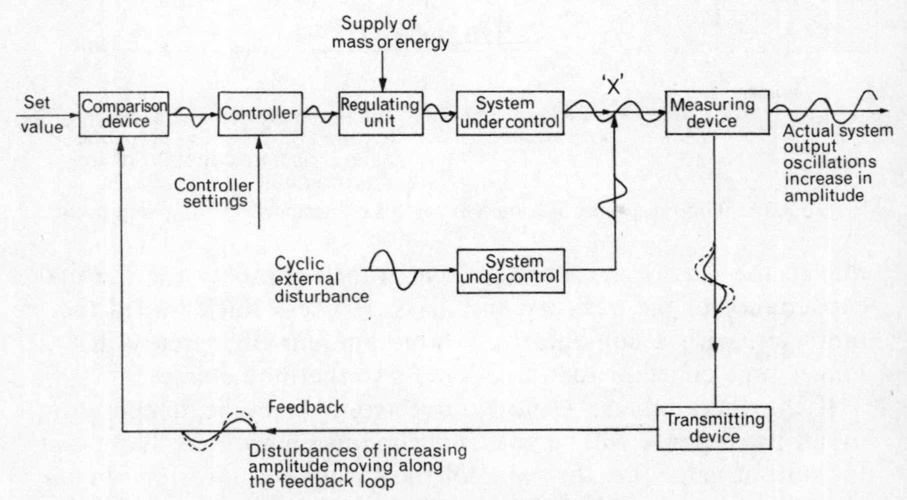

Figure 3.40 Unstable automatically-controlled system.

the input value by a certain phase angle, both of which increase with an increasing frequency of variation of the input. This type of relation between the input to, and the output from, a particular system block is called the *transfer function* of the block.

Many components of control systems have a transfer function similar in form to that of the thermometer, and some of these are shown in Figure 3.39.

Figure 3.40 represents a complete control system subjected to cyclic external disturbance, where variations in the measured variable occur at point X on the diagram. The external disturbance is modified by the system and imposes a variation of the system output at point X, which is measured and transmitted through the feedback loop to the comparison device. The comparison device produces an error signal, which is modified in the controller and sent on to the regulating unit, which alters the flow of mass of energy into the system. The system output then changes and combines with that produced by the external disturbance at X. Under certain circumstances, it is possible for the variations due to the external disturbance to combine with the variations in the system output due to the presence of the closed loop, so that a disturbance of greater amplitude than before moves along the feedback loop. Under these circumstances, the oscillations in the system output increase, and the system is said to be *unstable*.

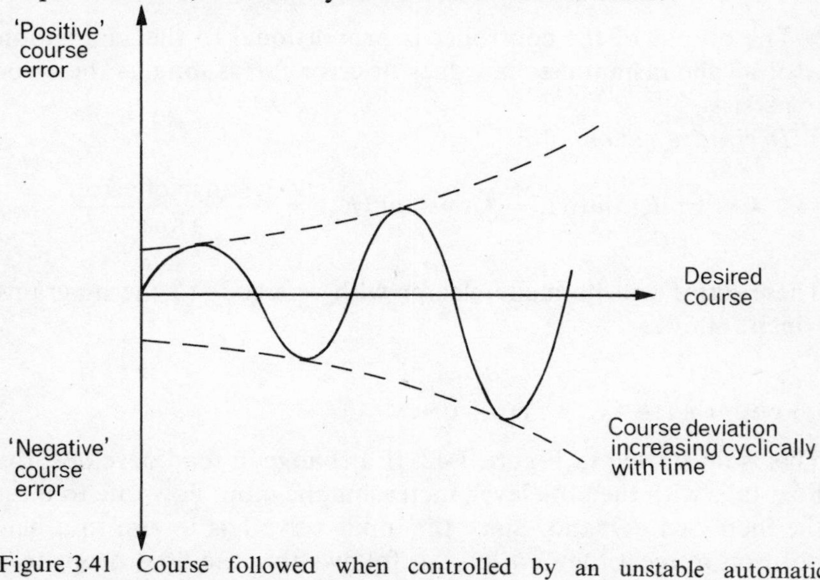

Figure 3.41 Course followed when controlled by an unstable automatic steering-system.

Using a practical example, imagine a series of waves imposing a cyclic external disturbance on a small vessel, causing the course of the vessel to fluctuate. If the automatic steering system were unstable, any rudder actions would add to the amplitude of the course variations. The course followed is illustrated in Figure 3.41.

CONTROLLER OUTPUTS

In order to modify system responses, a controller may use the error signal in such a way that the system performance and response are optimized. The output of the controller can be adjusted by various settings which alter the controller outputs. These outputs are:

1. *Proportional action:*

   $$\text{Controller output} = \text{Constant } (K_P) \times \text{Magnitude of error}$$

   The constant is called the *gain* of the controller, which is an amplification factor.
2. *Integral action:*

   $$\text{Controller output} = \text{Constant } (K_I) \times \sum (\text{Error} \times \text{Time})$$

   The output of the controller is proportional to the summation of all the instantaneous values of error for as long as the error persists.
3. *Derivative action:*

   $$\text{Controller output} = \text{Constant } (K_D) \times \frac{\text{Change of error}}{\text{Time}}$$

These terms will be made clearer with reference to the diagrams which follow.

(*a*) PROPORTIONAL ACTION ONLY

This is illustrated in Figure 3.42. If a change in load develops, the float falls with the tank level, increasing the input flow rate to meet the increased demand. Since the input valve has to rise to a new position to meet the new load, it follows that the float drops to a new position, which means that the level must drop below the

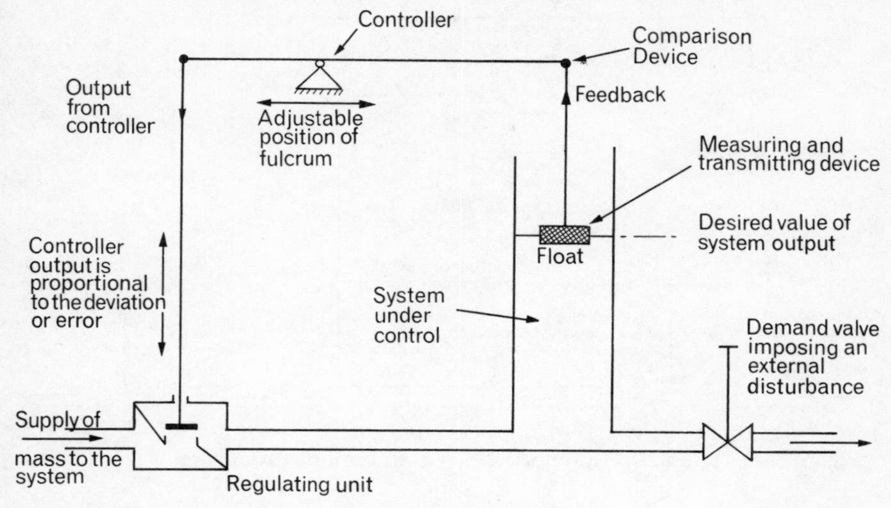

Figure 3.42 Example of proportional control action.

desired value. This droop, or offset, with change in load, is a characteristic of proportional-process control systems. The diagram shows that moving the fulcrum towards the float would reduce offset, since the same change in the position of the input valve would be obtained with a smaller variation in tank level. Moving the fulcrum to the right would also increase the gain of the controller. Hence it can be concluded that increasing the gain reduces offset. However, increasing the gain of the controller also increases the probability of instability in the system.

(*b*) PROPORTIONAL + INTEGRAL ACTION

It can be seen from Figure 3.43 that the addition of the integral output ensures that the controller produces an output to change the position of the supply-regulating valve, *so long as an error persists*. Hence integral control eliminates offset, but tends to encourage the instability of a system, especially when large errors are present at system start-up or when large changes in desired value are demanded.

(*c*) PROPORTIONAL + INTEGRAL + DERIVATIVE ACTION

A controller is called 'three-term' if it produces such an output. From Figure 3.44 the contribution of derivative action to the positioning of the supply-regulating valve can be assessed. If the

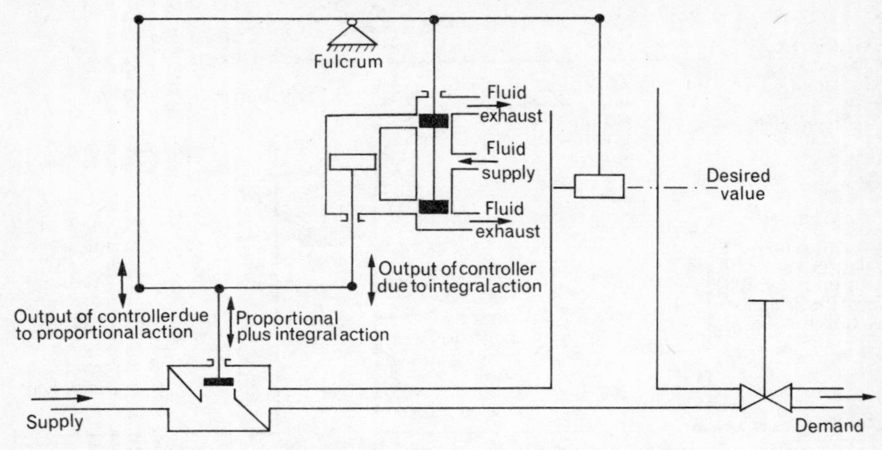

Figure 3.43 Example of proportional plus integral control action.

float falls rapidly, due to a large increase in load, the left-hand side of the lever rises quickly, allowing the piston contained in the cylinder to rise, bringing the cylinder with it. The rise in the cylinder makes its contribution to the increase of supply, by the link shown. Clearly, the cylinder is centred by its springs to its mid position, when the error is not changing. Therefore, derivative control reduces the large initial error which would otherwise result from a large increase in load. In addition, a derivative term tends to increase the stability of a controlled system, which may allow the proportional and integral term settings to be adjusted to give faster system response.

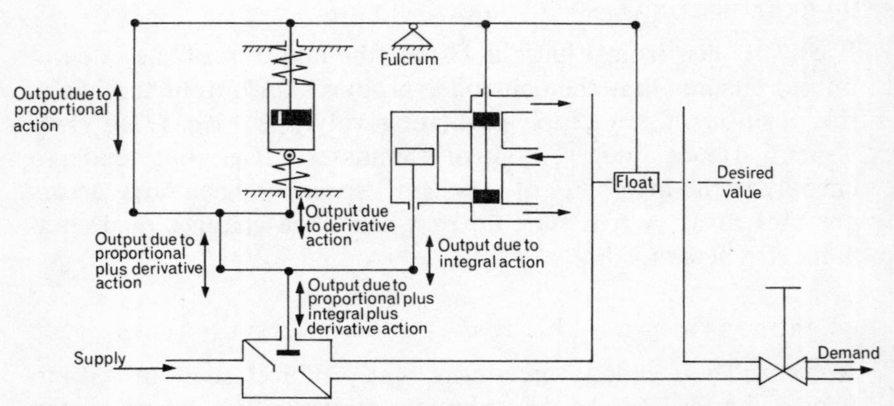

Figure 3.44 Example of proportional plus integral plus derivative control action.

The reader is reminded that this book is being written as an introduction to the subject. Hence the above treatment has ignored many other ways of modifying system response which can be examined in specialist books on the subject of automatic control.

In Chapter 4, some examples of shipboard systems will be examined, using the principles which have been introduced in this chapter.

Chapter 4

Shipboard Systems

INTEGRATED SHIP INSTRUMENTATION SYSTEMS

As the number of measuring devices increases, it becomes necessary to collect the information regarding the performance of each ship system in a central position, because control decisions made in one subsystem affect the performance of other systems.

For the power plant, information is collected in the machinery

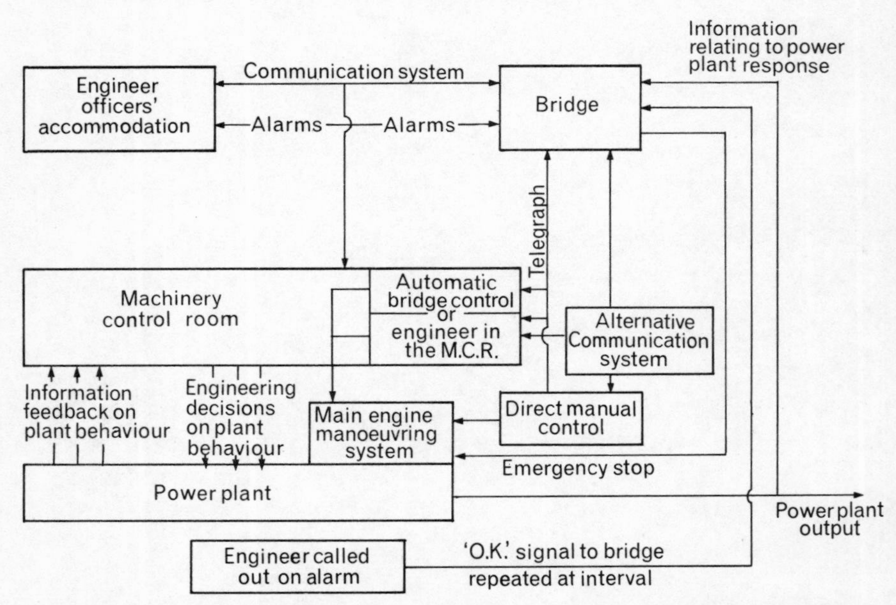

Figure 4.1 Information flows in a power plant system incorporating a machinery control room.

control room (M.C.R.). The machinery control room can thus become the complete operational centre for all the energy conversion systems described in Chapter 2, as indicated in Figure 4.1. For the cargo control system, a similar information-gathering centre may be established. The major difference in the operation of the centres is the fact that, while in power plant control it is usual to use many automatic closed-loop systems, the great majority of cargo control systems are manually operated, at present (in 1976).

The control rooms are designed ergonomically, particular attention being paid to the grouping of information on the display panels. For example, the pointers in a group of gauges showing normal operating conditions are all in the same relative position. By this means it is easy to spot a gauge indicating an abnormal condition.

MIMIC DIAGRAMS

Mimic diagrams, illustrating the major mass and energy flows in the appropriate system, are constructed as part of the control console. Miniature gauges, mounted in the surface of the panel, indicate important measuring devices at important points in the system. The position of control valves and pumps, and their operational state, are also indicated on the mimic or graphic display panels. The presence of these diagrams enables operational and diagnostic decisions to be reached by personnel much more quickly and safely than would be the case in their absence.

ADVANCED SYSTEMS

In some countries, originally France and latterly Japan, the M.C.R. has been combined with the bridge control position, and the bridge watchkeeper is in a position to make engineering decisions regarding the power plant, as well as navigational decisions. However, the production of these watchkeepers would demand radical changes in present British training patterns, which would reflect those developing elsewhere.

DATA-LOGGING SYSTEMS

Comprehensive data-recording and alarm systems, as shown in Figure 4.2, are often incorporated in ships fitted with centralized control rooms. These systems have two main functions:

1. *Alarm monitoring*. The values of certain measured variables are each scanned at short time intervals, and the reading obtained is compared with adjustable reference levels. If such a reading is outside its pre-set limits, audible and visual alarms are instigated, and a time-identified print-out of the alarm condition is given.

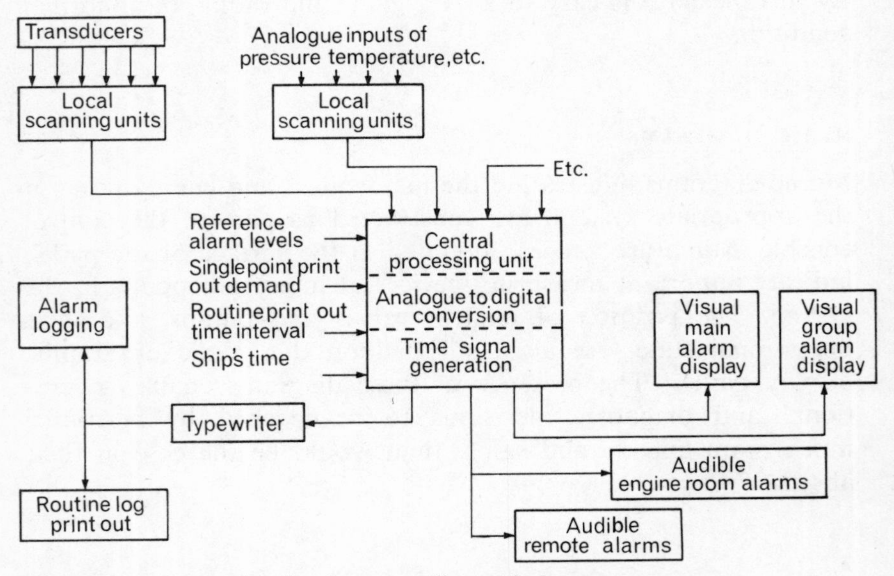

Figure 4.2 Comprehensive data-recording and alarm system.

2. *Data recording or data logging*. The measured value of the system variables can be printed on demand, or automatically, at certain set time intervals. The data so recorded may be analysed to assemble operational data, to provide the ship operators with the information they require to diagnose faults, to plan their maintenance schedules, and to provide the shore staff responsible for the specification and design of the vessel with feedback of operational data.

PRESENTATION OF INFORMATION

Most transducers currently present information in analogue form. For example, the position of the pointer on a voltmeter is analogous to voltage; the output of a thermocouple is analogous to the temperature difference between the hot and cold junctions; the positioning of the hands on a clock are analogous to elapsed time.

Analogue measurement, display, and recording systems represent a physical variable, according to the *continuous* measurement of that variable in time. *Digital* measurement, display, and recording systems represent the value of a measured variable at *discrete* time intervals, in some form of code, which may be a train of pulses, a set of perforations on a card, or a set of figures.

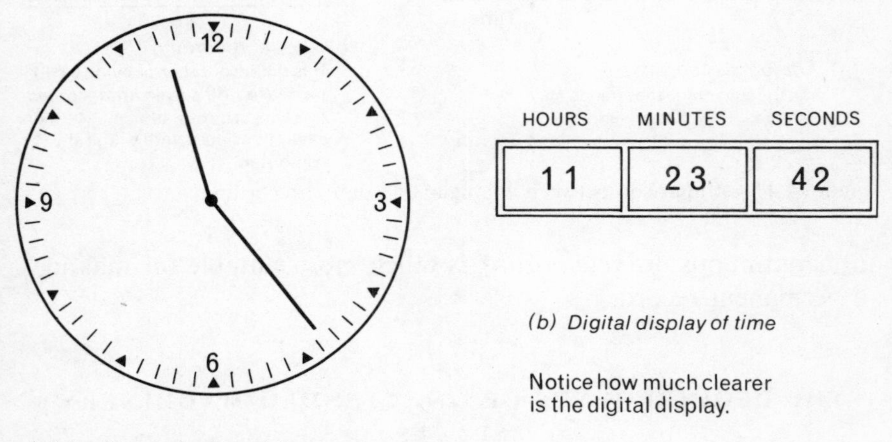

Figure 4.3 Comparison between analogue and digital display.

Figure 4.3 further explains the difference between the analogue and digital presentation of information. Notice the discrete nature of the digital time display; i.e. there is a discrete interval of time between the display of 11 hours, 23 minutes, and 42 seconds, and of 11 hours, 23 minutes, and 43 seconds. The time interval could be reduced to 1/100 or 1/1 000 of a second, but it would still be discrete.

Figure 4.4 illustrates the difference between analogue and digital recording. Notice that, although the analogue record enables a trend diagnosis to be easily made, the digital record, because of its

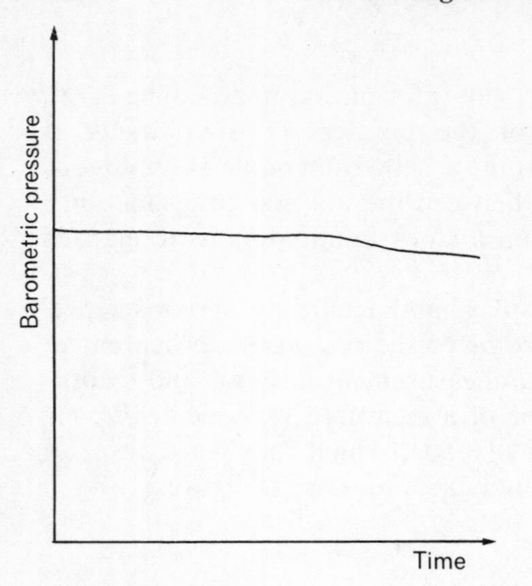

TIME	BAROMETRIC PRESSURE
10 20	1003
10 40	1003
11 00	1004
11 20	1004
11 40	1005
12 00	1005
12 20	1004
12 40	1004
13 00	1004
13 20	1004
13 40	1003
14 00	1003
14 20	1002
14 40	1002
15 00	1002
15 20	1002

(*a*) *Analogue recording*
has the advantage that trends may be more easily diagnosed.

(*b*) *Digital recording*
has the advantages of being easily processed, displayed and recorded; this type of information may be more easily interfaced with a digital computer.

Figure 4.4 Comparison between analogue and digital recording.

noncontinuous discrete nature, is much more suitable for making a permanent record.

THE REQUIREMENTS OF UNATTENDED MACHINERY SPACES

Unattended machinery spaces have been made possible due to developments in measurement and control systems. In order to sustain the safe operation of the ship under U.M.S. conditions, the following principal requirements should be met.

BRIDGE POSITION

1. A control system must be provided to operate the main engine.
2. A reliable means must be provided to stop the main engine should the control system fail.
3. Alarms are necessary to indicate a failure in the energy supply to the control system.

4. Two methods of communication must be provided between the bridge and engine room, one of which is independent of the main energy supply.
5. A manual method of manoeuvring the engine must be provided in the engine room.
6. The following instrumentation is usually fitted at the bridge position:

   (*a*) Propeller speed.
   (*b*) Direction of propeller rotation or pitch position.
   (*c*) Starting air pressure available (for diesel engines).
   (*d*) Indication of engine malfunction.

ALARM SYSTEM REQUIREMENTS

If a ship is to be operated under U.M.S. conditions, a comprehensive alarm system must be installed, with the following principal requirements.

1. Machinery faults must be indicated at a main control station.
2. Engineering personnel must be made aware that such a fault has occurred.
3. If the ship is operating with U.M.S., the bridge watchkeeper must be made aware:

   (*a*) that a fault has occurred;
   (*b*) that the engine room staff are aware of the fault and are attending to it;
   (*c*) that the fault has been rectified.

4. The bridge watchkeeper must be used to safeguard the safety of the person in the engine room, since normally he will be unaccompanied.
5. An effective system must be provided to enable the engineer called to the engine room by an alarm, to ask for assistance from other engineers.

PROTECTION OF THE POWER PLANT

This requires the engine to be stopped automatically should a serious fault condition occur, e.g. a complete loss of lubricating oil pressure. The main engine will slow down automatically, or alternatively be slowed down by the bridge watchkeeper, for certain other faults, which depend on the plant installed. Some installations have an override facility on the bridge, which may be used if the continued supply of motive power is absolutely necessary for the safety of the vessel.

SUPPLY AND DISTRIBUTION OF ELECTRICAL ENERGY

It is necessary to provide essential lighting and power should the normal supply fail. Hence, a system which automatically starts up a standby generator and connects it to the electrical distribution system is required.

Hence, the essential safety features for U.M.S. operation can be summarized as:

1. Bridge control of main engines.
2. Comprehensive fault detection and alarm systems.
3. Reliable fire detection and alarm systems.
4. Alternative sources of electrical energy.

These will now be considered in more detail.

BRIDGE CONTROL SYSTEMS

Bridge control systems for different types of power plant are shown in block diagram form below.

BRIDGE CONTROL OF STEAM TURBINES

Figure 4.5 illustrates the manual control of a steam turbine. Should an astern manoeuvre be ordered under 'full away' conditions, the engineer must prepare the turbine for manoeuvring. Certain essential services maintained by steam extracted from the turbine may have to be transferred; the ahead manoeuvring valve can be closed; the astern guardian valve and various drains are opened. Astern steam can then be admitted at the appropriate rate, and the turbine will stop and accelerate in the astern direction.

Bridge control replaces the preparatory work of the engineer officer by logic circuits, and his rate of steam admission by programmes specially developed for a particular installation. A bridge control system is illustrated in Figure 4.6, and the safety circuits,

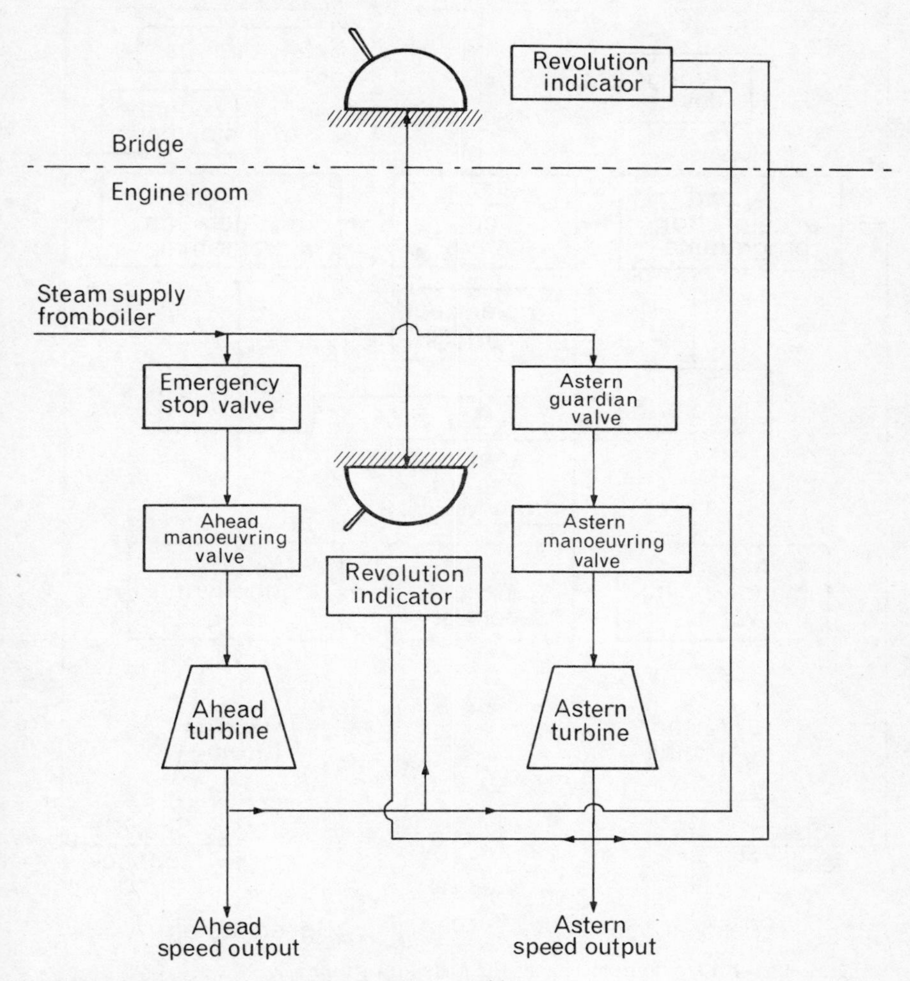

Figure 4.5 Manual control of a steam turbine.

with their effect on the availability of power, are shown in Figure 4.7. An idling device is fitted to prevent rotor distortion should it be held at stop for more than a few minutes. Sometimes the idling device can be overridden, e.g. to prevent mooring ropes fouling the propeller, but in such a case the response of the turbine to the next movement may be limited.

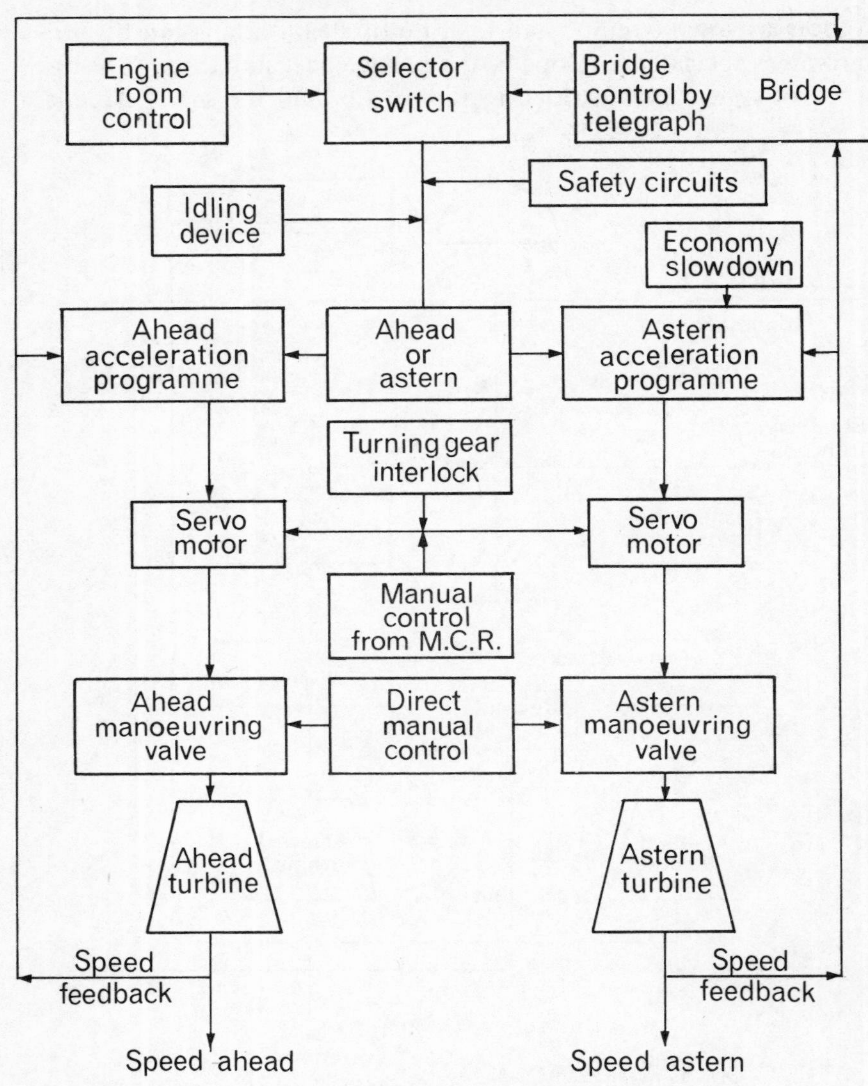

Figure 4.6 Bridge control system for a steam turbine.

BRIDGE CONTROL OF A DIRECT-REVERSING DIESEL ENGINE

In Chapter 2 the manual operation of such an engine was described. Again, in the automatic system (*see* Figure 4.8), various logic circuits and acceleration programmes replace the operational skills of the engineer officer. The programme should prevent the engine from running within its critical revolution range. Safety

FAULT	EFFECT
Loss of lubricating oil Loss of electrical power	The turbine shuts down, and is protected from water torque acting on the propeller
High condenser level High or low lubricating oil temperature Excessive vibration Excessive rotor movement High or low boiler water level Excessive torque Excessive thrust	Turbine slows down

Figure 4.7 Safety circuits for a steam turbine.

circuits are incorporated so that the engine will stop for failure of lubricating oil pressure, and slow down for:

1. High scavenge-air temperature.
2. Crankcase oil mist.

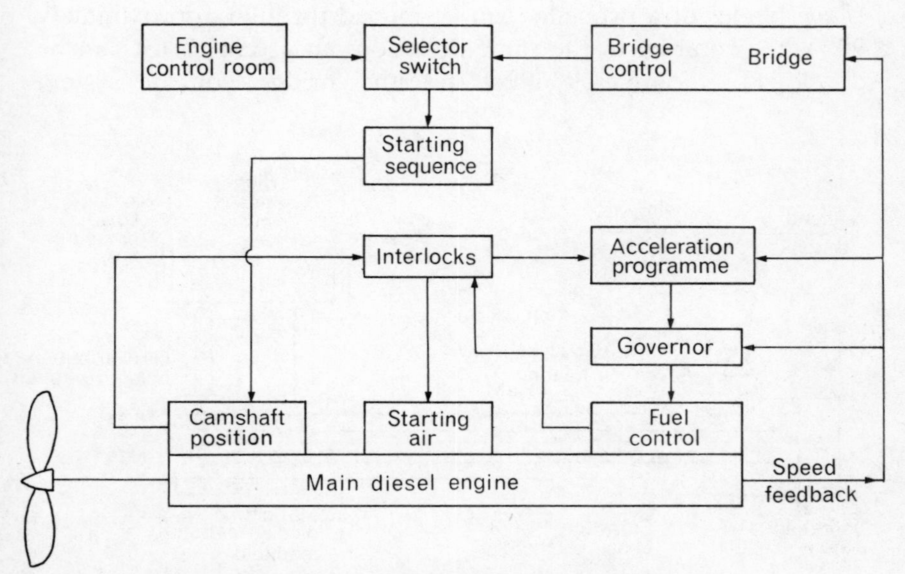

Figure 4.8 Bridge control of a direct-reversing diesel engine.

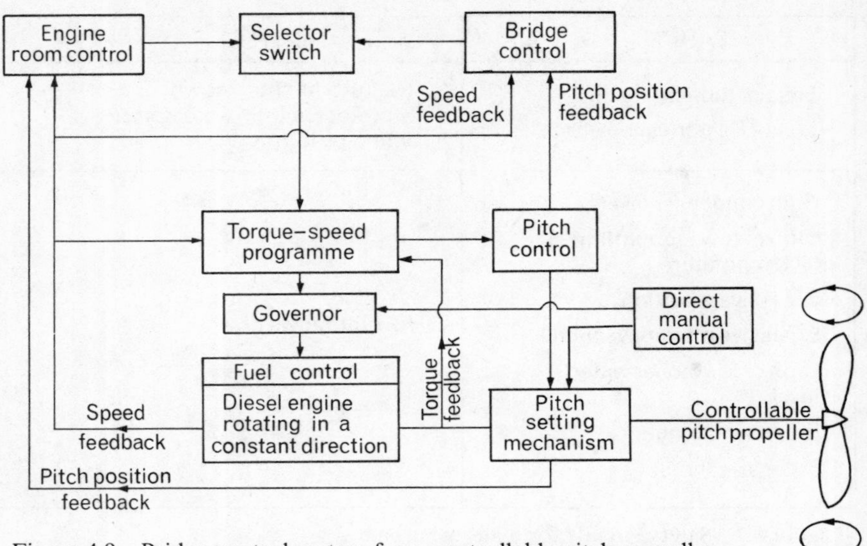

Figure 4.9 Bridge control system for a controllable pitch propeller.

3. Low piston-coolant flow.
4. Low coolant pressure.

BRIDGE CONTROL SYSTEM FOR A CONTROLLABLE PITCH PROPELLER

If the blades of a propeller can be moved through approximately 90° while rotating in the same direction, an astern thrust can be developed. Figure 4.9 illustrates the bridge control system

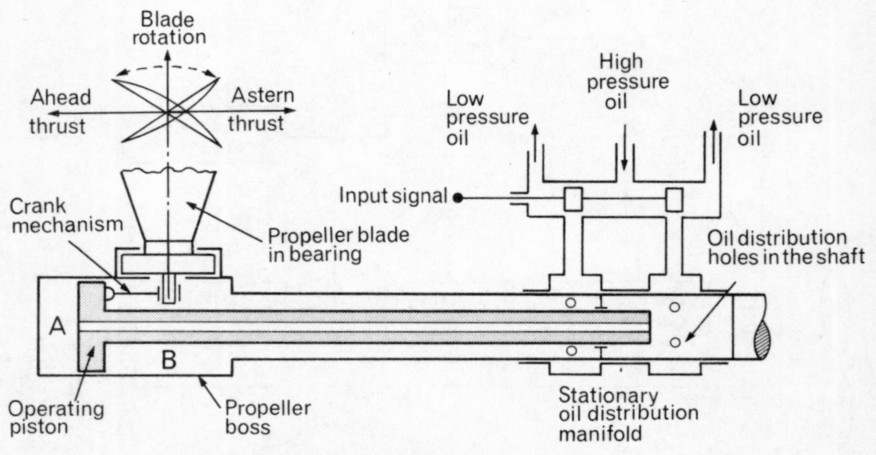

Figure 4.10 Controllable pitch propeller.

dedicated to the control of propeller blade angle, and hence of propeller pitch, and to the control of engine speed and torque.

A diagrammatic sketch of a controllable pitch propeller is shown in Figure 4.10. The position of the oil distribution control valve can be altered by the bridge control system, and, in conjunction with the control of engine speed, the desired thrust in the appropriate direction can be obtained by allowing high-pressure oil to flow into space A or B as required.

COMPREHENSIVE FAULT DETECTION AND ALARM SYSTEMS

Systems suitable for U.M.S. operation, as well as for manned operation, have been described earlier in this chapter under 'Data-Logging Systems' and 'The Requirements of Unattended Machinery Spaces'.

FIRE DETECTION AND ALARM SYSTEMS

Detectors, e.g. the rate of temperature rise detector, were described in Chapter 3. A circuit diagram, illustrating a method of using these detectors to instigate an alarm, is shown in Figure 4.11.

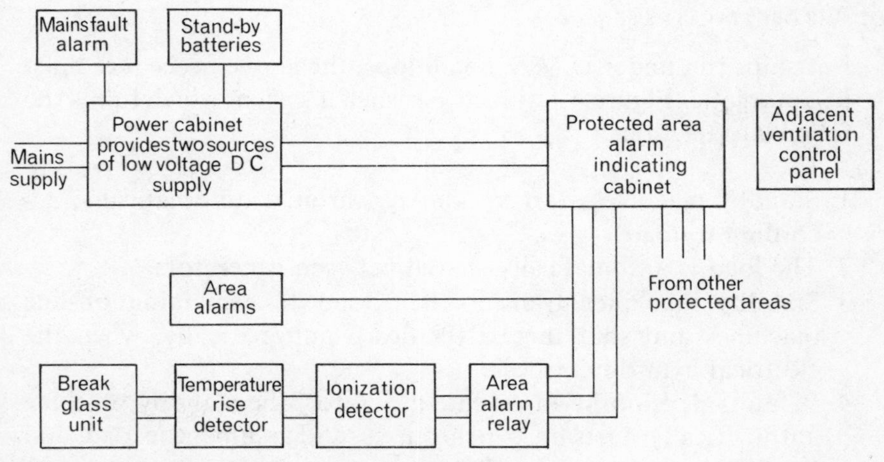

Figure 4.11 Fire detection and alarm system.

ALTERNATIVE SOURCES OF ELECTRICAL ENERGY

The reliability of the electrical supply is essential to maintaining the controllability of all but the smallest vessels.

Synchronization of Alternators

MANUAL SYSTEM

In a manual system, should a fault develop in an on-line generator, or the load increase to an unacceptable level for the on-line plant, the engineer officer can start up standby generators and transfer some or all of the electrical load to the incoming generator. For an alternating-current three-phase distribution system, he must follow this procedure:

1. Adjust the voltage of the incoming alternator to suit the system voltage.
2. Ensure that the voltage vectors of the system and of the incoming machine are in phase.
3. Adjust the frequency of the incoming machine to suit the existing distribution-system frequency.
4. Close the circuit breaker when these conditions are satisfied.
5. Balance the distribution of electrical load between machines.

AUTOMATIC SYSTEM

For ships run under U.M.S. conditions, the above procedure must be automatic. Figure 4.12 illustrates such a system which fulfils the following functions:

1. Standby machines start up and synchronize automatically, according to load.
2. The load is automatically shared between generators.
3. Standby machines transfer their load to remaining on-line machines and shut themselves down automatically, when the electrical load diminishes.
4. If faults develop in an on-line generator, the standby machine automatically starts up, synchronizes, and assumes the load, and the faulty machine shuts down.

5. Should the system fail to operate, and for certain faults in the prime mover, an alarm is instigated.

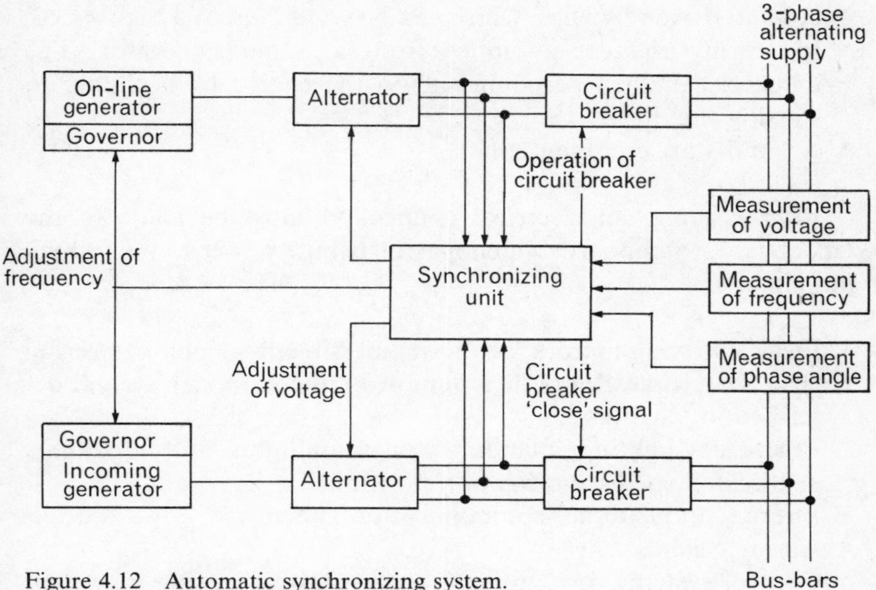

Figure 4.12 Automatic synchronizing system.

Emergency Generators

Emergency generators may be required to supply electrical energy, should the main generation system fail to maintain the supply of electrical energy. They are specified for ships carrying above a certain number of passengers, and for ships whose emergency electrical power requirements are above the capacity of storage batteries.

Usually, the emergency generators are diesel driven and have their own self-contained supply of superior grade fuel. They may be air cooled or may use a radiator and fan. They are located in a safe and easily accessible position above the bulkhead deck, which must be adequately ventilated.

For starting the diesel, a store of energy is required, which may be provided by:

1. Batteries.
2. An air reservoir in which the pressure is automatically maintained.

3. An hydraulic reservoir which can be pumped up by hand. The stored energy can then be released to operate a starting device.
4. A hand-driven flywheel which can be wound up to a high speed, so storing the energy input from a human operator. The flywheel can then be connected to the engine by a clutch, so starting the diesel.
5. A hand-start cranking handle.

After starting, an electrical connection must be made to the emergency switchboard, which may distribute power to the following circuits:

1. Electric driving motors for the submersible bilge pump, steering gear, watertight doors, fire pumps, or the sprinkler system described in Chapter 5.
2. Emergency lighting, including navigation lights, and floodlights at the lifeboat disembarkation positions.
3. Internal and external communication systems.
4. Alarm systems.
5. Control systems. It is difficult to re-establish engine power on many modern ships, without the aid of the relevant control systems. On these ships the emergency generator supplies power to these systems.

An emergency generator is not usually required to operate in parallel with the main generators. In these cases the main generator supplying the emergency circuits is automatically disconnected at the emergency switchboard, as the emergency generator is connected to the emergency circuits. However, in some other ships the emergency generator is integrated with the main electrical-distribution system and must therefore be paralleled in the same way as an incoming main generator.

INSTRUMENTATION FOR HULL SURVEILLANCE

The wave-induced motions and stresses in a ship's structure are very complex. In most ships, it is left for the captain to use his intuition and experience in assessing the effect of different weather conditions on the hull, but systems have now been developed which replace the intuitive approach and monitor the stresses set up in the hull in a seaway.

Strain gauges are used to measure stress at the points indicated in Figure 4.13. A linear accelerometer measures vertical accelerations in the bow, and a pressure transducer is used to detect

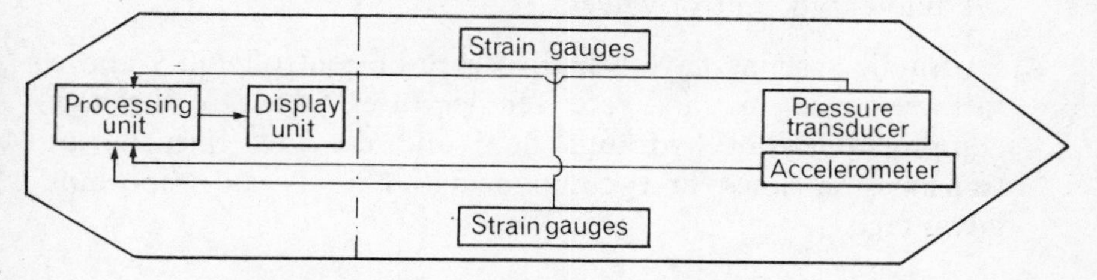

The processing unit statistically analyses the readings obtained and computes a standard deviation from the static values. These are displayed on the bridge.

Figure 4.13 Hull surveillance system.

the onset of 'slamming' under the fore part of the vessel. An alarm and display system is fitted at the bridge position, and the officer of the watch can alter the ship's speed or heading, or both, to reduce the stresses and motion of the vessel to acceptable limits.

HYDRAULIC SYSTEMS

Hydraulic Power Transmission

Hydraulic power transmission is finding an increasing application on board ship, because of the following major advantages:

1. The transmitting and receiving hardware is small in size.
2. There is no inherent time delay in preparing the receiving hardware for use. Contrast the preparation time for a steam and hydraulic windlass on a V.L.C.C., where getting steam to the windlass can take half a morning.
3. An hydraulic system can be used to fulfil many operating requirements. The central hydraulic-power unit can supply fluid under pressure to operate:

   (*a*) fire and ballast pumps;
   (*b*) fixed and portable submersible cargo-pumps;

(*c*) bow and stern thrust units;
(*d*) winches and windlasses;
(*e*) hatch covers;
(*f*) watertight doors;
(*g*) stern and bow ramps for roll-on roll-off ferries;
(*h*) fluid-cargo control valves.

4. Hydraulic systems have a high degree of controllability and a fast response in moderate temperatures, due to the low compressibility of hydraulic fluid and the low elasticity of transmission lines and components. This is an important advantage.

Against these advantages must be set the fire risk associated with the containment of a high-pressure inflammable fluid, and the large pressure drops which occur in exposed hydraulic lines in cold weather conditions.

HYDRAULIC PUMPS AND MOTORS

These are usually of the rotary positive-displacement type. Figure 4.14 illustrates how one such type works; the fluid flows indicated show how the device may be used either as a pump or, with the flows reversed, as a motor.

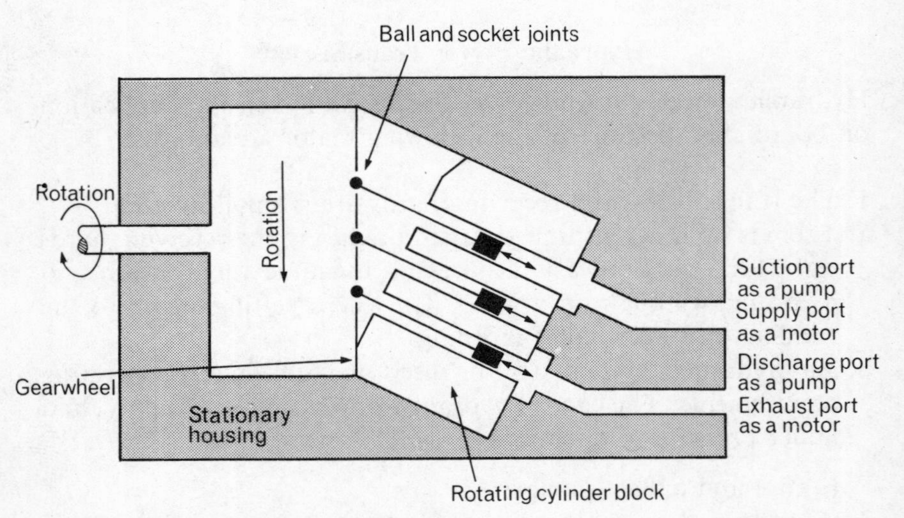

Figure 4.14 Hydraulic motor or pump.

VALVE ACTUATORS

Valve actuators or regulating units receive an input signal from either the controller, in an automatically controlled system, or the operator, in a remotely-controlled nonautomatic system.

The *power* required by the valve actuator may be supplied in one of three ways:

1. *Electrically*. A remotely operated motor is used to drive a valve spindle through a gearing system.
2. *Pneumatically*. An example is shown in Figure 4.15. The control

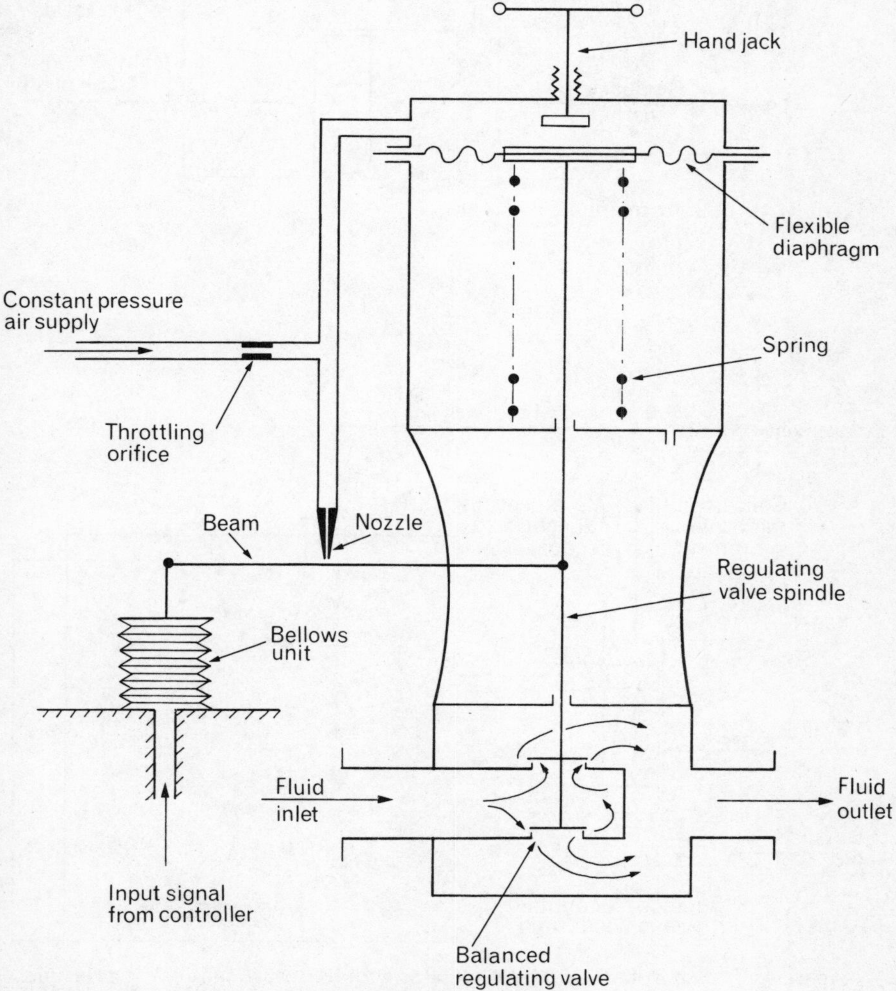

Figure 4.15 Pneumatically-actuated regulating valve.

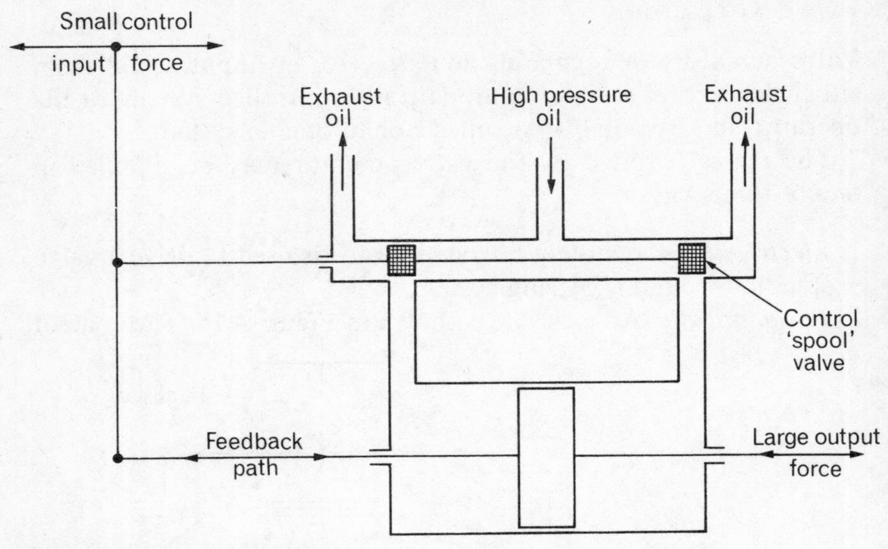

Figure 4.16 Linear hydraulic actuator.

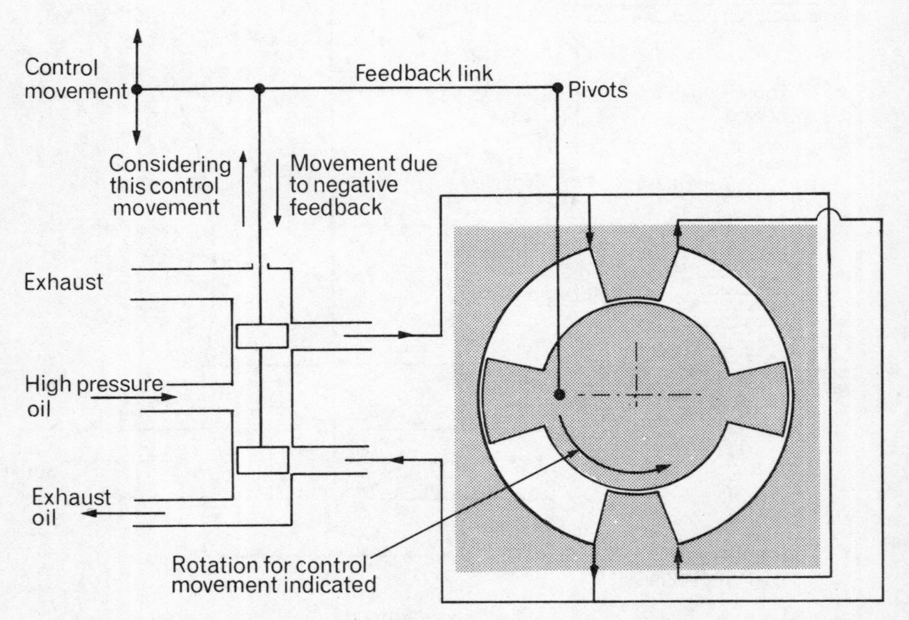

Figure 4.17 Semirotary hydraulic actuator with feedback (suitable, in principle, for a steering gear).

signal is fed to the bellows, which alter the clearance between the nozzle and beam. If the control signal increases, the back pressure behind the nozzle also increases, which causes the valve to close. As the valve closes it moves the beam downwards, providing the negative feedback which tends to re-establish the original beam–nozzle clearance, so that the valve comes to rest in the position appropriate to the magnitude of the control signal.

3. *Hydraulically.* Figures 4.16 and 4.17 illustrate the operation of:

   (*a*) a linear hydraulic actuator,
   (*b*) a semirotary hydraulic actuator.

   The position of the control valve is changed by the input control signal, which allows oil to flow through passages to operate the actuator. As the actuator moves, its movement is transmitted along the feedback path, so cancelling out the original movement of the control valve. Thus, the output movement of the actuator is proportional to the input control movement.

Fluid-Cargo Handling Systems

Fluid-cargo handling systems have been developed because:

1. The power required to operate large valves in a reasonable time has exceeded the power available from manual operators.
2. The operational demands of the cargo-handling process require that information relating to the system performance be collected in a control position and be properly displayed with the aid of mimic diagrams, so that safe and rapid operational decisions can be made.

Figure 4.18 illustrates a typical system for the remote operation of cargo control valves, using a type of linear actuator. Notice that no automatic valve-position feedback is employed. The operator relies on the feedback of information relating to the position of the valve, to implement control decisions.

To close the valve, the pressure line is connected to the top and the bottom of the actuator. Since the effective area of the top of the piston is greater than that of the bottom, because of the area of the piston rod, the downward force exceeds the upward force, and the valve will close. A requirement of a classification society is that

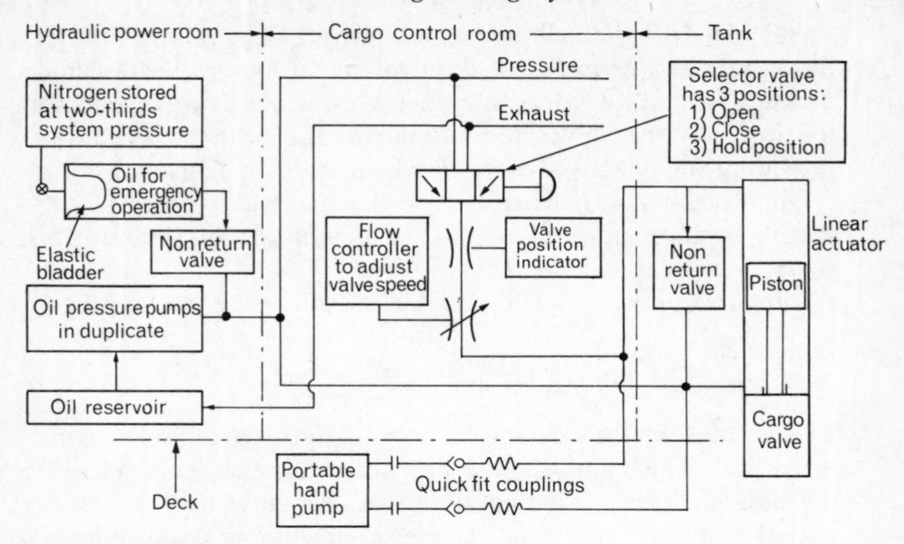

Figure 4.18 Cargo control system.

pressure failure should not cause the valve to open. From the diagram it can be seen that, should there be a pressure failure, any forces generated in the actuator will tend to keep the valve closed.

To open the valve, the exhaust line is connected to the top of the piston, and the pressure line to the bottom, so opening the valve.

To hold the valve, the exit of fluid from the top of the piston is closed by the selector valve. If the pressure of the hydraulic fluid in the space above the piston rises, due perhaps to the heat gained by the fluid from the cargo, the nonreturn valve relieves the excess.

EMERGENCY OPERATION OF VALVES

It must be possible to operate valves in spite of electrical power failure, or accidental or storm damage to the hydraulic pipework. Figure 4.18 illustrates two methods which would be incorporated in a typical system.

1. By releasing the energy stored in high-pressure nitrogen, enough power is available to operate four large valves, or a greater number of smaller valves, in approximately fifteen seconds.
2. Mobile hand-operated hydraulic pumps can be wheeled about to join the system by the snap connectors shown. Alternatively, a number of fixed hand-operated pumps can be fitted, and con-

nected to the appropriate valve by long flexible hoses. Shut-off valves are fitted, so that each section of the system can be isolated to allow for any necessary maintenance.

If possible, before cargo operations begin after a lengthy voyage, each cargo valve should be operated for a few seconds in each direction. If the vessel is loading through very long shore-lines, the energy stored in this column of moving oil is very great. Should a cargo valve be closed very quickly, the shock wave instigated is likely to damage the cargo valves. In such cases, the loading supervisor must give advice on minimum valve-closing times.

Ship-Steering Systems

The steering system may be operated under manual, or automatic, closed-loop conditions.

MANUAL CONTROL

Figure 4.19 illustrates the manual operation of an electro-hydraulic steering gear. The operator visually assesses the difference between the desired course and the actual course indicated.

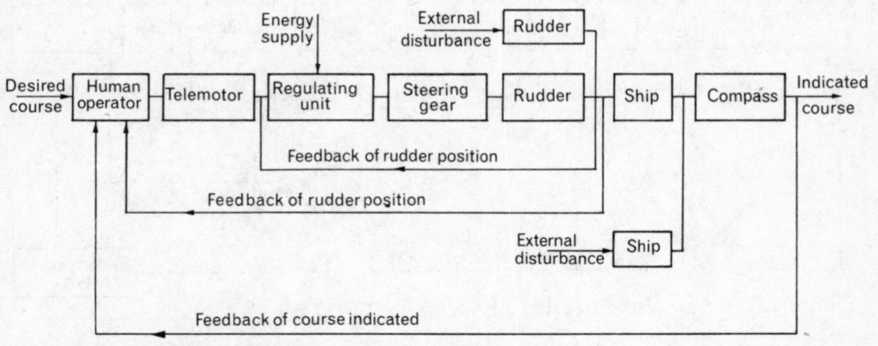

Figure 4.19 Nonautomatic steering of a ship.

Using the steering wheel and telemotor system, he passes a control signal to the variable delivery pump, which delivers oil to the rudder-actuating mechanism in order to reduce the course error. The operator can observe the rudder position, and when the ship's head begins to swing towards the desired course he can return the rudder to amidships, or even apply 'counter rudder' in the opposite direction, so that the ship, due to its large inertia, comes to rest pointing in the correct direction without overshoot.

The feedback loop within the rudder-operating mechanism is necessary for two reasons:

1. To check on the performance of the components within the steering gear.
2. To compensate automatically for undesired rudder movements. To protect the components of the steering gear, relief valves are fitted to allow the rudder to move if a large force is applied to it, e.g. by a heavy beam sea. It can be seen from the diagram that such a movement would be transmitted through the feedback lever, causing the variable delivery pump to re-establish the original rudder position without the intervention of the bridge watchkeeper.

AUTOMATIC CONTROL

The automatic closed-loop system replaces the skill and experience of the operator by automatic devices, as indicated in Figure 4.20.

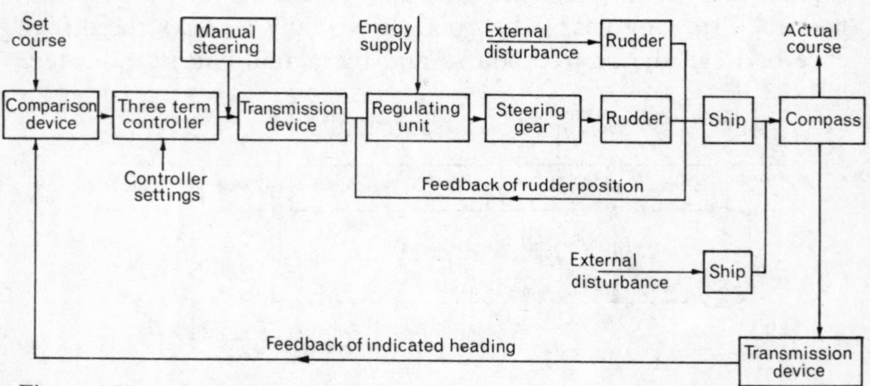

Figure 4.20 Automatic steering of a ship.

A *three-term controller*, and the control actions it produces, was described in Chapter 3. The three-term controller in a steering gear system may generate its control actions electrically, but their definition and functional effects are exactly the same as those described earlier, using the level controller as an example.

1. The *proportional term* of the controller transmits a control signal to the steering gear, which is proportional to the course

error. The ratio of control action to error signal can be altered, and is usually called the 'rudder multiplier'.

2. The *derivative term* of the controller produces an output when the course of the vessel is changing. The 'counter rudder control' measures the rate at which the vessel is swinging back on to course, and generates a control action which applies opposite rudder and so prevents overshoot. The time constant associated with this control action can be altered for different conditions of loading.
3. The *integral term* of the controller produces an output for as long as a course error persists. With all aft ships especially, the centre of pressure of the superstructure is aft of the centre of pressure of the wetted hull surface. With a beam wind, a couple is generated which tends to turn the ship into the wind. The rudder position required to combat this is called 'weather helm', which can be generated in either of two ways:

   (*a*) *manual weather helm*, which can be set on the steering system and which creates an artificial midships position for the rudder, or
   (*b*) *auto weather helm*—the value of the steady controller output due to integral action is proportional to the required weather helm, and this output is used to set the rudder to the appropriate weather angle automatically.

Three *other possible control settings* are as follows:

1. *Damping control* introduces a time delay into the controller output, so that the steering gear does not respond to repetitive course errors, which are due to ship yaw as it meets a cross sea. This saves wear and power consumption in the steering system.
2. *Sheering control* establishes a 'dead band' in the controller, so that the steering gear does not respond to small random course errors, which may in fact be generated within the control system itself.
3. *Alarms*, generated by a steady off-course signal by large rudder movements, or by power failure, are usually fitted to the system.

Ship Stabilization Systems

Ships roll, pitch, heave, yaw, sway, and surge. Of these motions, only roll can be dealt with effectively in practice, by using, gen-

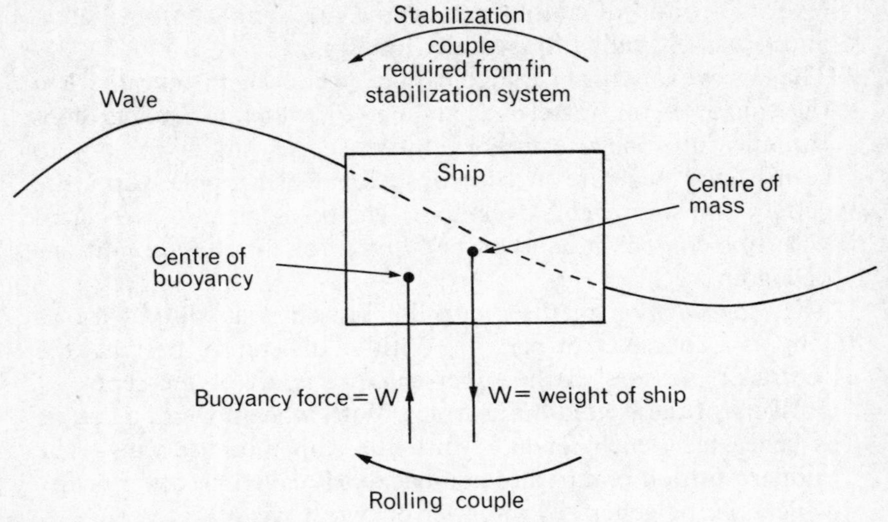

Figure 4.21 Cause of rolling.

erally, a fin stabilization system. This generates a stabilizing couple to oppose the wave-induced roll couple, as indicated in Figure 4.21. The force in the fin, required to generate the stabilizing couple, is generated by altering the angle of incidence of the fins to the water passing over them.

The stabilizing couple must be carefully controlled to optimize the behaviour of the ship. Figure 4.22 illustrates a typical ship stabilization system. The measurement system uses gyroscopes to

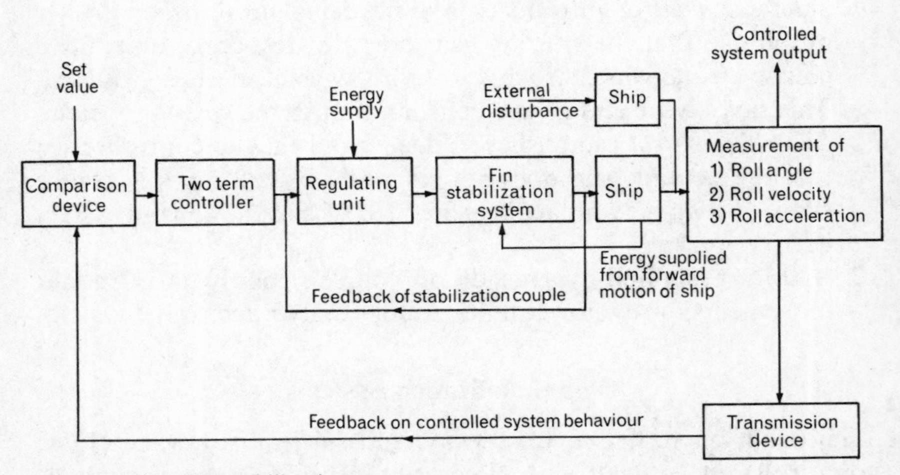

Figure 4.22 Fin stabilization system.

measure the roll angle, velocity, and acceleration of the ship. An error signal is generated which is usually processed by a proportional and derivative controller. An integral action term is usually omitted, since such action would cause the fins to attempt continually to correct a list angle, which may have been caused by the consumption of fuel or by passengers' crowding to one side of the deck. Such fin action would impose an unacceptable drag on the vessel.

THE SHIP'S TELEGRAPH

The ship's telegraph is an information-transmitting and -receiving system. It may use a high-output synchro system, similar to that described in Chapter 3, in the manner shown in Figure 4.23. Other systems may use a faceplate commutator to produce differing voltages, which can operate a receiver.

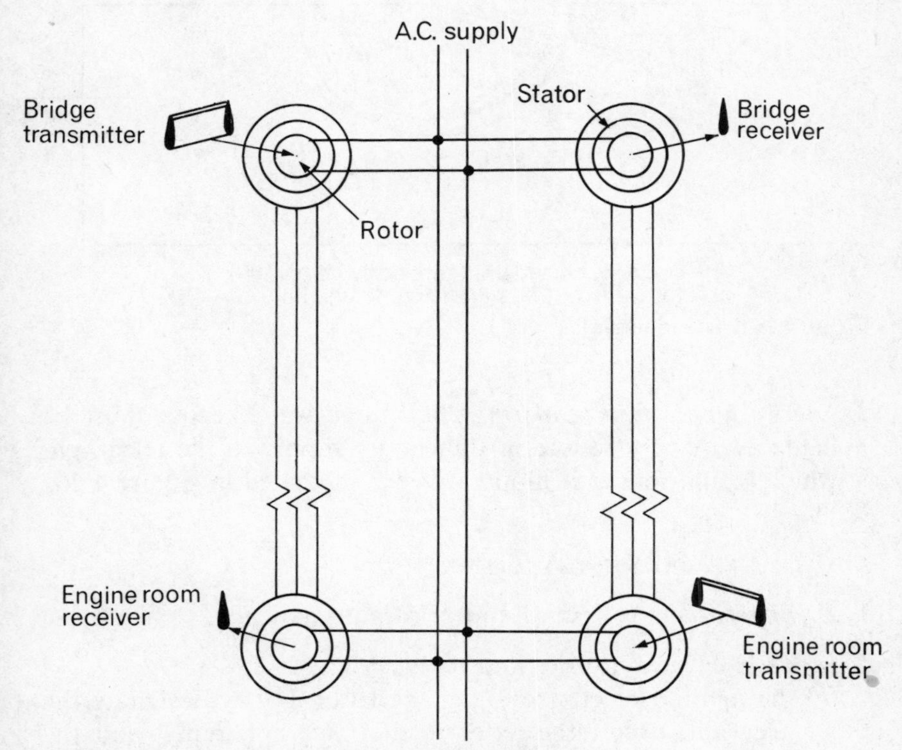

Figure 4.23 Ship's telegraph using a synchro system.

BRIDGE ALARMS

The bridge control position must also be fitted with three alarms:

1. *Attention alarm.* Figure 4.24 shows such a system. If the transmitting handle is in a position different from that of the receiving pointer, a current flows, instigating an alarm to the appropriate receiver.
2. *Wrong way alarm.* This sounds when the direction of engine rotation does not correspond to the position of the telegraph lever. Such a system is shown in Figure 4.25.

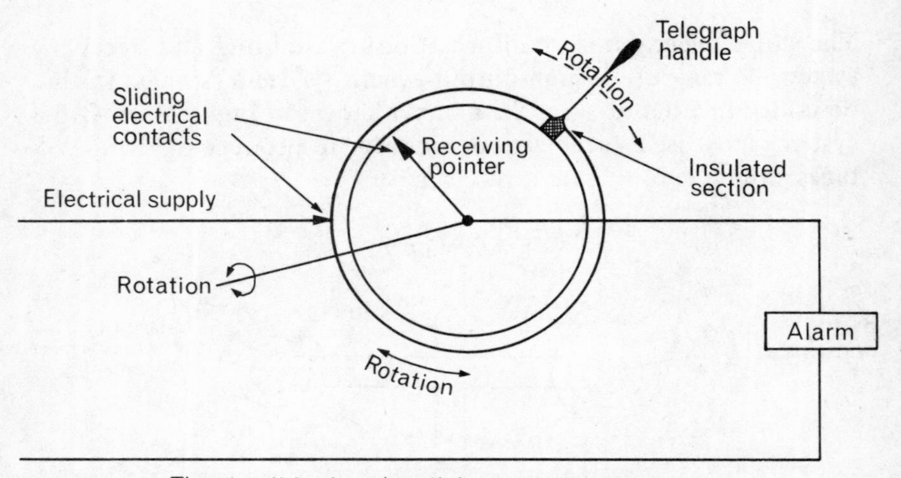

Figure 4.24 Attention alarm.

3. *Energy-supply failure alarm.* The bridge watchkeeper must be made aware of a failure in the energy supply to the telegraph, which would render it inoperative, as indicated in Figure 4.26.

ADDITIONAL BRIDGE FACILITIES

1. *Event recorders* to record and print automatically:

   (*a*) each telegraph order and reply;
   (*b*) the method of control—e.g. the letter T may designate the telegraph order, the letter R the telegraph reply, and the letter B that the engine is on bridge control;

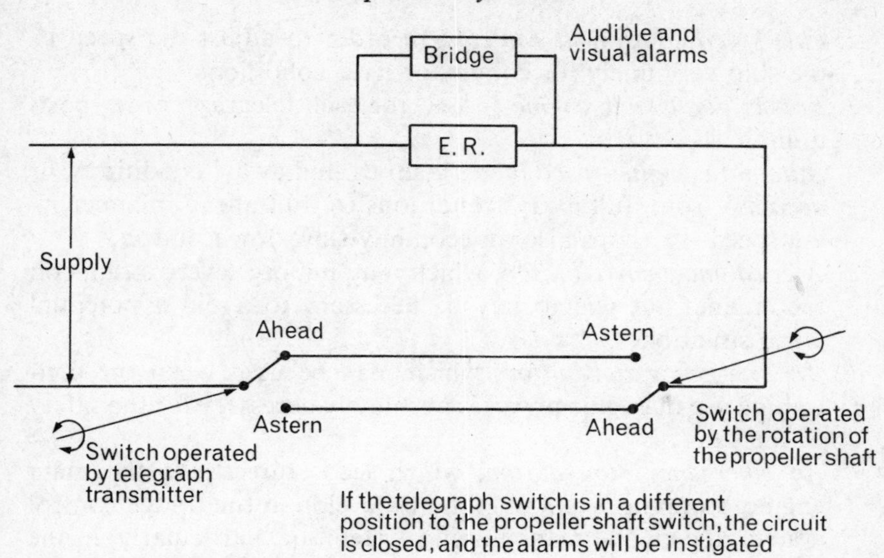

Figure 4.25 Wrong-way alarm.

(*c*) the day of the year, the hour of the day, and the minute of the hour, divided into ten-second intervals;

(*d*) the position of a controllable pitch propeller, or the engine revolutions ahead and astern, at the time the order was given.

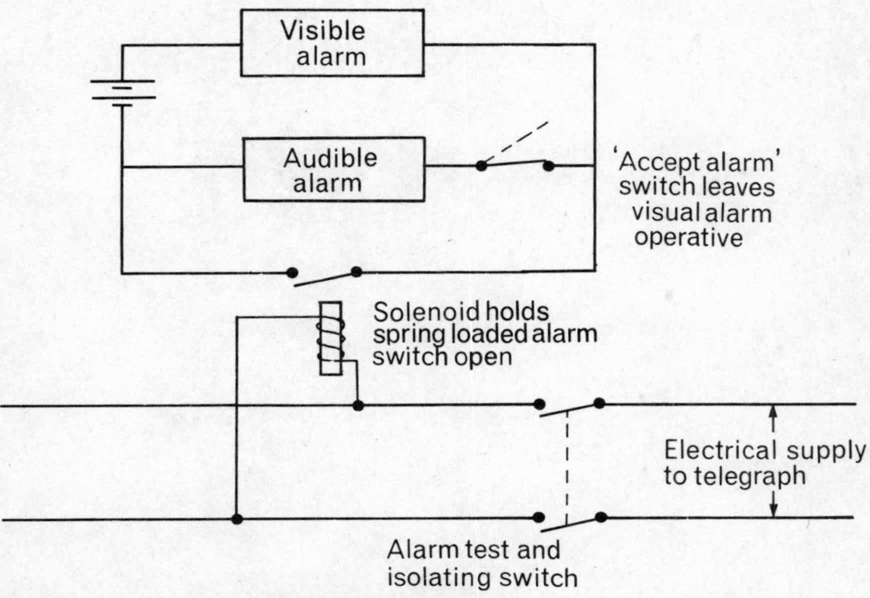

Figure 4.26 Alarm for loss of energy supply.

2. *Fine speed-adjustment controls*, in order to adjust the speed of the ship very finely for convoy or river conditions.
3. *Pre-set, adjustable engine speeds*, for each telegraph order position.
4. *Automatic engine-speed increase* up to 'full away' conditions, *or decrease* from 'full away' conditions to 'full ahead' manoeuvring speed—perhaps also an economy slow-down button.
5. *A crash manoeuvre button*, which may impose severe strains on the engine, but which may be necessary to avoid a potential crash situation.
6. *An emergency run button*, which may be used when the continuing use of engine power is absolutely necessary for the safety of the ship.
7. *An emergency stop button*, which leads directly to the main engine, stopping it, should a fault develop in the bridge control system, the engine manoeuvring system, or, particularly, in the pitch-setting mechanism of a controllable pitch propeller.

Chapter 5

Fire-fighting Systems

The loss of lives and ships due to explosion and fire continues at an unacceptably high level.

Clearly, any person claiming to be competent must study this topic carefully, and be prepared to demonstrate his competence to an examiner. The large, fixed, fire-fighting systems on board ship may be fairly complex engineering installations, and the staff of the engineering department are examined rigorously on the mechanical and operational details of these plants.

However, in many ships today, in particular those on which the engineering staff has been reduced due to the application of the U.M.S. concept, it is quite possible for the total engineering complement to be incapacitated together, e.g. by an engine room explosion. The safety of the ship and its personnel may then rest solely on the ability of the deck officers to operate the fixed installations.

In many companies today, a standing order exists requiring that a deck officer joining a ship be made thoroughly conversant with the fixed fire-fighting installations, the remotely-operated oil pumps and valves, and all the engine-room ventilation systems, by a senior member of the engineering department. This book confines its attention only to those systems specified in the Syllabus for Master (Foreign Going), published by the Department of Trade.

Fire-fighting systems can be subdivided broadly into fire prevention, fire detection, and fire fighting.

PREVENTION OF FIRE AND EXPLOSION

Flammability Limits

The essential co-requisites for an explosion and fire are:

1. Inflammable vapours and materials.
2. Oxygen, mixed with the vapour and combustible material in the correct proportions, within a flammability range.
3. A source of ignition.

Dealing specifically with hydrocarbon vapours, Figure 5.1 illustrates the relative proportions of hydrocarbon vapour and oxygen which are the pre-requisites to a fire and explosion. From the

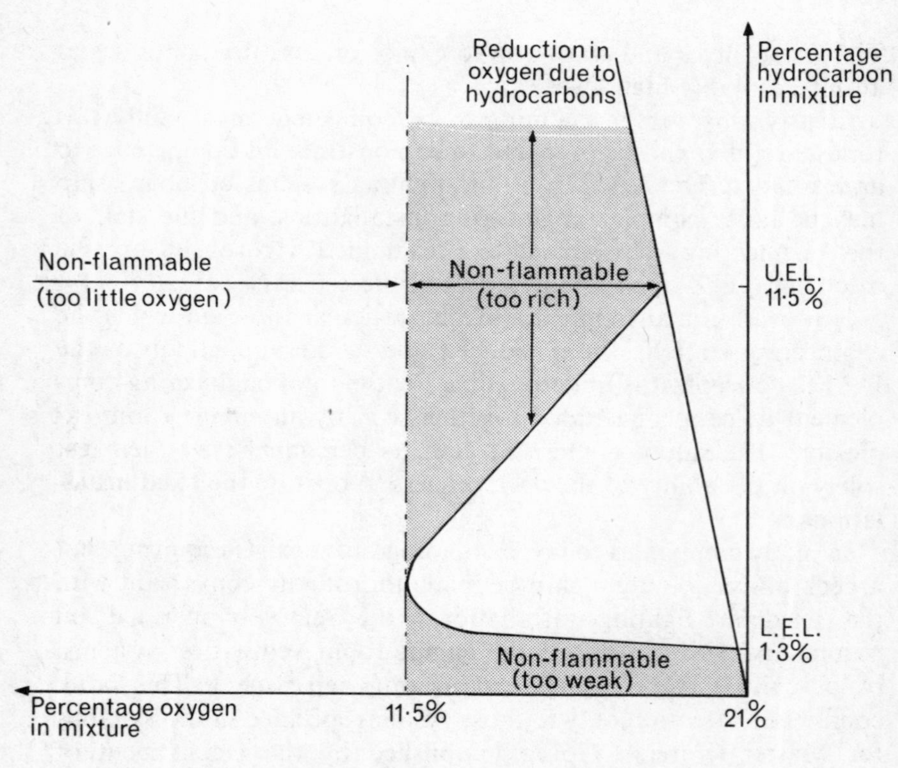

Figure 5.1 Flammability of hydrocarbon/nitrogen/oxygen mixtures.

diagram it can be seen that reducing the oxygen content below a certain value will 'inert' the tank atmosphere and drastically reduce the probability of fire and explosion in the inerted space.

Inert-Gas Protection System

In Chapter 2 the combustion process was discussed, and an analysis was given of the flue gas which may be obtained from the efficient combustion of fuel oil. Examination of the analysis in Figure 2.14 shows that, if the flue gas could be cleaned and cooled, the following gas proportions could be expected:

Carbon dioxide	12·0–14·5%
Oxygen	2·5–4·5%
Sulphur dioxide	0·02–0·03%
Nitrogen	about 77%
Remainder	water vapour and solids

Clearly, this cool mixture, composed largely of the inert gases carbon dioxide and nitrogen, would be effective in preventing explosion if it were pumped into a tank to replace the original oxygen in the tank. Such a system is illustrated in Figure 5.2.

The volume rate of gas flow produced by the system is at

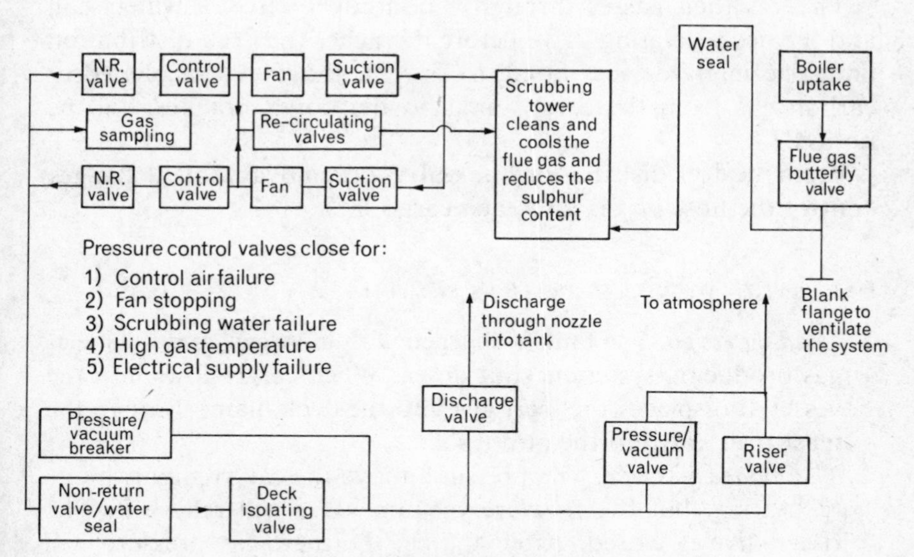

Figure 5.2 Inert-gas protection system for oil tankers.

least 1·25 times the total volume-throughput rating of the cargo pumps.

Flue gas is drawn through butterfly valves and a pipe loop which extends above the top of the scrubbing unit, designed to prevent a backflow of water into the uptakes. The gas enters the scrubbing unit and is subjected to a spray of water, which condenses out the steam in the flue gas and dissolves most of the sulphur dioxide. The gas then passes to the fans, which may be fitted with a recirculating gas line back to the scrubbing unit. This gas line allows for:

1. Starting up the system before the pressure control valves to the deck line are opened.
2. Reducing pressure fluctuations when both fans are in operation.
3. Avoiding overheating of the fans when the demand for gas is low.

The fans discharge through control valves which maintain the pressure in the gas main at a desired value, which may vary between 250 mm and 1 000 mm of water head.

A gas sampling point leads to an oxygen analyser, which instigates an alarm when approximately 8 per cent of oxygen is reached. The instrument is bypassed when gas freeing.

The gas then passes through a nonreturn valve, a water seal, and the deck-isolating valve before it reaches the deck distribution line. The many devices fitted, to prevent the backflow of hydrocarbon gas from the cargo tanks to the boiler uptakes, will be noticed.

From the deck distribution line, valves or removable blank flanges control the flow of gas into each cargo tank.

OPERATION OF THE INERT-GAS SYSTEM

1. *Loading cargo*. The tank is charged with inert gas, and the inert-gas-producing system is shut down. When cargo flows into the vessel, it displaces the inert gas into the deck main, through the mast riser valve to the atmosphere.
2. *The loaded passage*. The pressure above the cargo may rise due to heating, but the pressure/vacuum valve will relieve it. The riser valve is closed. After a time, the inert gas pressure will probably fall. If the pressure drops to about 100 mm head of

water, the system is operated until the pressure reaches about 750 mm head of water.

3. *Discharging cargo*. Inert gas is discharged to the tanks at a high rate; the riser valve is closed. The pressure–flow characteristics of the fan are such that the tank cannot be overpressurized if the discharge suddenly ceases. Should the inert gas system fail, the liquid pressure and vacuum breaker supplements the pressure/vacuum valve on the riser, if this is necessary to protect the tank structure from a high vacuum.
4. *Ballasting*. During ballasting, the system is operated as when loading cargo, unless the ship has facilities for taking in dirty ballast while discharging. If this is the case, the system is left operational and the mast-head vent valve is adjusted to maintain the pressure in the vapour main.
5. *Tank washing*. Prior to washing, the tank is purged with flue gas. During tank washing, the inert gas system produces a high flow rate in order to maintain a moderate tank pressure. Certain tank orifices may be open.

COMBINED FIRE-DETECTION AND FIRE-FIGHTING SYSTEMS

Inert Gas Generators

It has been shown above how flue gases can be used to provide a supply of inert gas to prevent explosion and fire. In cargo ships, inert gas may also be used to fight fires in the cargo holds of the vessel. Since the availability of adequate flue gases from the main engines of a cargo ship cannot be guaranteed in port, a separate system is fitted which supplies clean, cool, inert gas and distributes it to the holds. Such a system is shown in Figure 5.3.

A good quality fuel oil is burned in the combustion chamber. The flue gas, after passing through the scrubbing unit, may have the following composition:

Carbon dioxide	14–15%
Oxygen	0–1%
Nitrogen	approximately 85%

The volume rate of gas flow produced must be enough to fill the largest cargo compartment in four hours.

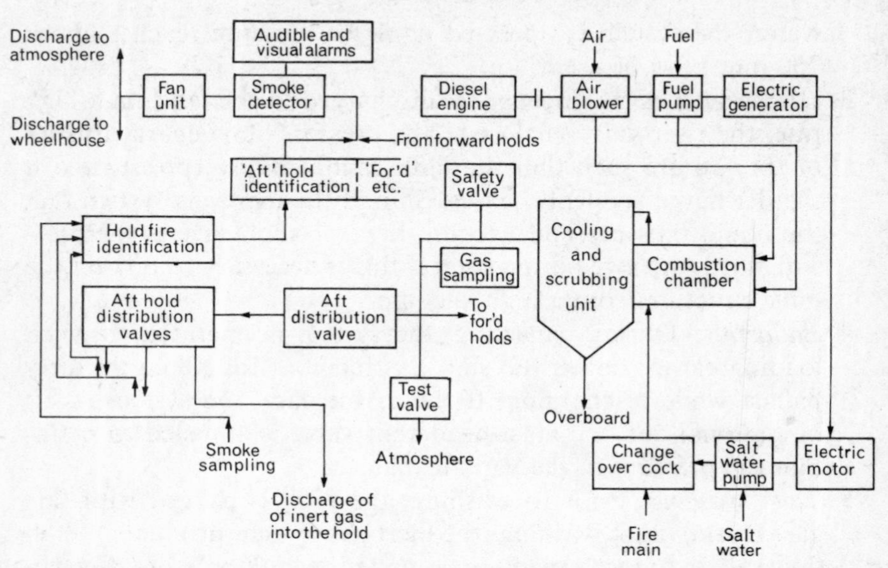

Figure 5.3 Inert-gas fire-detection and fire-fighting system for a cargo ship.

A sample of the atmosphere in each hold is drawn through the sampling system, which leads to a smoke-detecting photo cell. This cell instigates a visible and audible alarm if smoke is detected. The hold affected is then identified and the ventilation or humidity

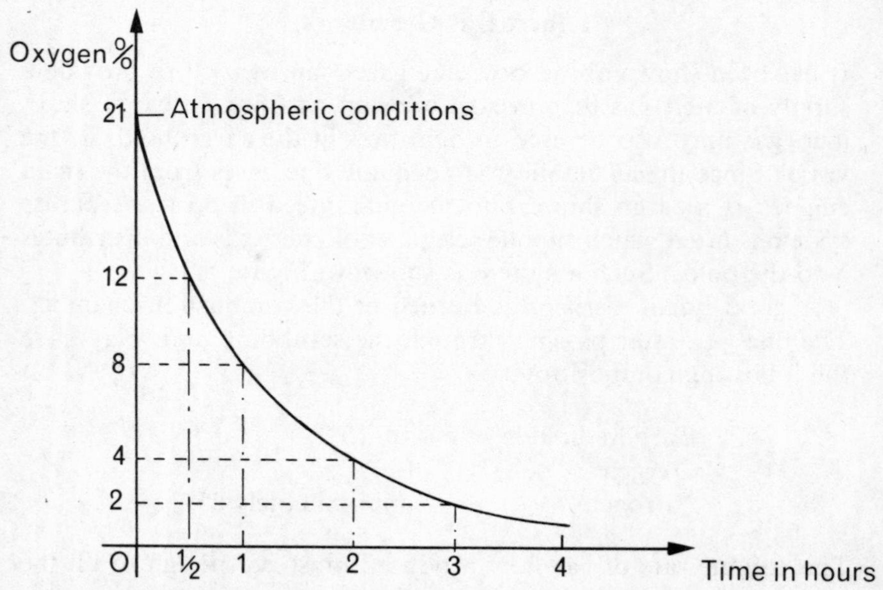

Figure 5.4 Reduction of oxygen content in a cargo hold over time.

control system is secured. The inert gas generator is then started up, and inert gas is pumped into the hold. The graph in Figure 5.4 shows the reduction of oxygen in the hold, on a time base.

The inert gas generator can be run on test, using the discharge valve shown. When the holds are empty, the line through to each hold can be checked by blowing air through the pipe, and the smoke detection system can be checked by holding a smouldering rag beneath the smoke sampling points.

Carbon Dioxide Systems

In these systems, a store of the inert gas, carbon dioxide, is used to displace the oxygen in a protected space. Such a system is shown in Figure 5.5.

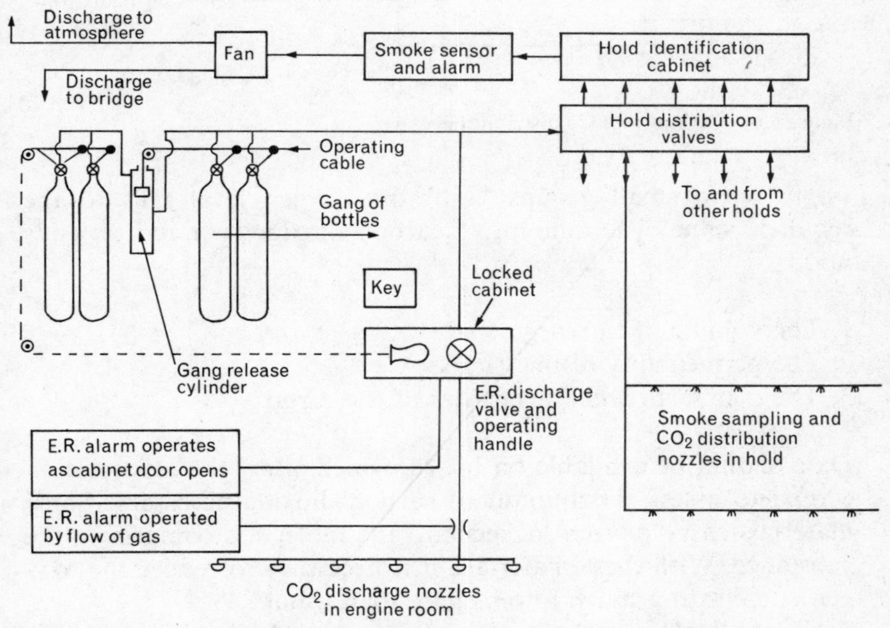

Figure 5.5 Fire-detection and fire-fighting system using carbon dioxide.

CARGO SPACES

When the double seated valve, shown in Figure 5.6, is screwed down, a sample of the air in the cargo space can be drawn through the distribution pipes into the smoke detection system. Photo-electric and ionization devices may be used to detect the presence of combustion particles and to instigate audible and visible alarms. The hold identification cabinet may then be inspected and the affected cargo space identified.

The space is prepared for receiving the gas in a manner similar to that employed by the inert gas system.

The double seated distribution valve is opened and carbon dioxide is gradually discharged into the space, from bottles opened

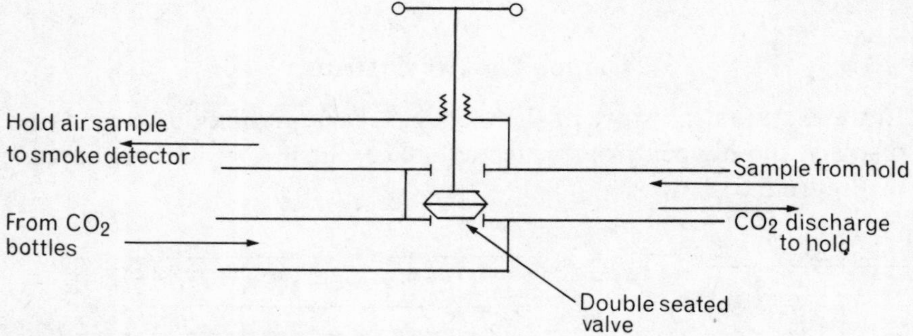

Figure 5.6 Carbon dioxide distribution valve.

singly or in small groups, until the oxygen level falls to the required value. The amount of carbon dioxide required depends upon:

1. The volume of the cargo space.
2. The permeability of the cargo.
3. The combustion characteristics of the cargo.

Data should be available on board, which allow the officers concerned to assess the amount of carbon dioxide necessary. Some materials carry oxygen locked into the molecules comprising the substance. With these materials, it is necessary to reduce the oxygen content to a much lower figure than usual.

Checks can be made on the system when the holds are empty. The line through to each hold can be checked by blowing air through it. A suitable connection for this may be provided near the distribution valve. The smoke-detecting unit can be tested by holding a smouldering rag beneath a sampling point.

Flow indicators, e.g. a small propeller, are fitted at the outlet end of the smoke-detecting pipe in the hold identification cabinet. These permit a visual check that the smoke detection system is in operation, provided that the integrity of the distribution pipes has been proved.

The mass of gas in the bottles can be checked by:

1. Weighing.
2. Using a level-detecting unit, which consists of a source of radioactive particles and a geiger counter. It is moved up and down the bottle, and the counter changes its reading as the level of liquid carbon dioxide in the bottle is reached.

MACHINERY SPACES

Fires in engine rooms tend to develop much more quickly than most fires in cargo spaces, because of the large quantities of hot oil present. Therefore, to extinguish a machinery space fire the carbon dioxide must be released very quickly, once the decision to evacuate the machinery space has been reached.

Before releasing the gas, all personnel are evacuated, all skylights and ventilators are closed, and all fans, oil pumps, and oil valves are shut off by the remote controls, some of which were indicated in Figures 2.16 and 2.17.

The key for the cabinet housing the engine-room discharge valve is kept in a glass-fronted box adjacent to a locked cabinet (*see* Figure 5.5). When the locked cabinet is opened alarms sound in the machinery spaces, indicating the imminent discharge of gas. The discharge valve is opened and the operating handle is pulled.

The operating handle opens two bottles, which discharge gas into the gang release cylinder. A piston within this cylinder can thus develop enough power to open the complete gang of bottles together, very quickly. The contents of the operating bottles are not wasted; they can enter the distribution system as the piston completes its stroke. This system meets the requirement that enough carbon dioxide must be discharged into a machinery space to fill approximately 30 per cent of its volume in a two-minute period.

Sprinkler System

The basic idea of the sprinkler system is shown in Figure 5.7. If a fire breaks out in the protected space, heat rises, melting the nylon

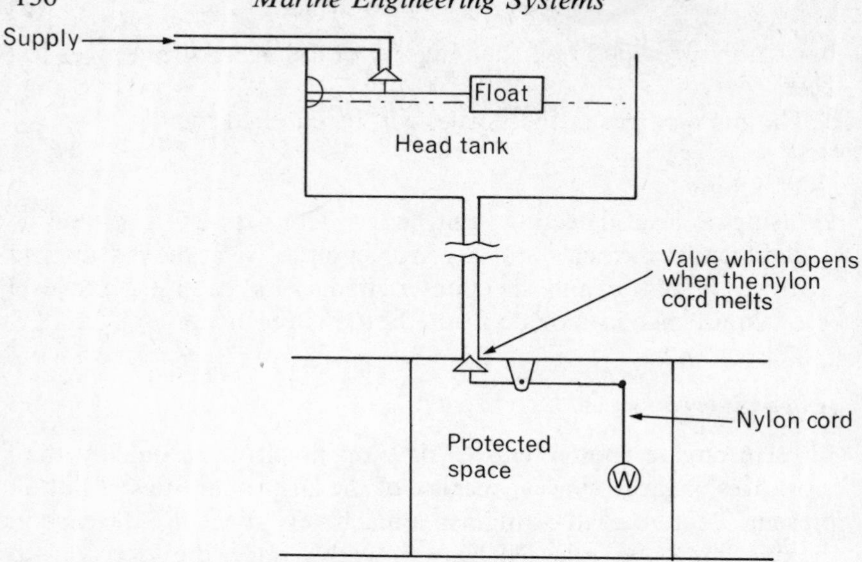

Figure 5.7 Basic idea of the sprinkler system.

cord at a predictable temperature. The valve may then open, discharging water into the space and putting the fire out.

In a real shipboard system, the 'sprinkler head' replaces the nylon cord. The valve is kept closed by the presence of the glass bulb, which contains liquid, and a bubble of the vapour associated with the liquid. As the temperature of the bulb increases and the pressure gradually rises, the vapour is gradually absorbed by the liquid until eventually it is completely absorbed (*see* Figure 5.8).

Any further heating of the bulb results in a sharp pressure rise,

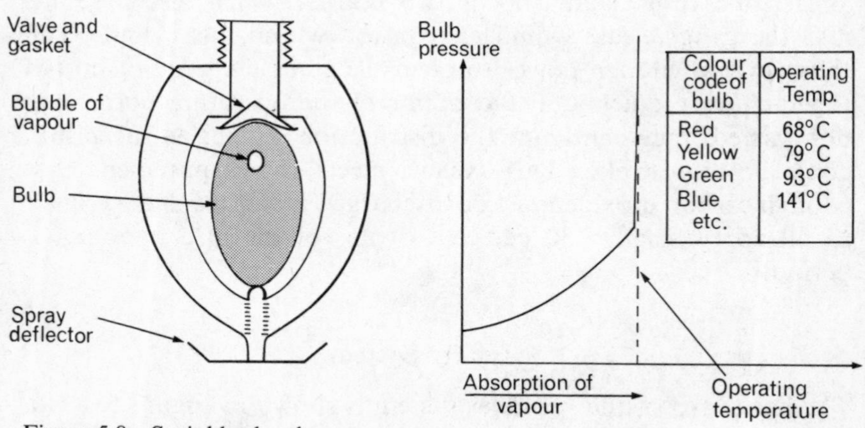

Colour code of bulb	Operating Temp.
Red	68°C
Yellow	79°C
Green	93°C
Blue	141°C
etc.	

Figure 5.8 Sprinkler head.

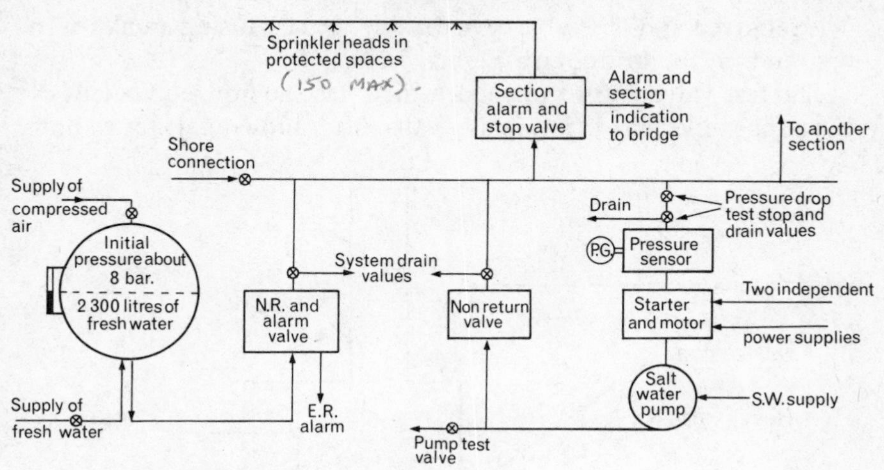

Figure 5.9 Sprinkler system.

since the liquid is virtually incompressible and expands at a greater rate than the glass bulb. The glass therefore shatters, at a predetermined temperature indicated by the colour of the bulb, allowing water to discharge into the protected space.

Events then proceed as follows (*see* Figure 5.9):

1. The air in the storage tank expands, forcing fresh water through the engine-room alarm valve, through the section valve, and into the protected space. The diagram of the section master valve (Figure 5.10) shows how an alarm is instigated on the bridge and an indication is given of the section affected.
2. As the air in the tank expands its pressure falls, and after a fall in pressure of about 3 bar the salt water pump automatically starts, maintaining a flow of water into the protected space. (If the fire-fighting party are very efficient, it may be possible to shut off the section master valve before the salt water pump cuts in.)
3. After the fire has been put out, the system is drained, the sprinkler head replaced, the system filled with fresh water, and the air pressure in the tank raised to its operational pressure of about 8 bar.

REGULAR TESTS ON THE SYSTEM

Regular tests on the system must be maintained, as follows:

1. In a definite time-identified order, each section valve is visited.
2. The leather strap securing the section valve in the open position

is removed and the valve closed a few turns, causing an alarm to sound on the bridge (*see* Figure 5.10).

3. The test valve is then opened, which has the flow equivalent of one sprinkler head. The heavy valve lifts, allowing water to flow

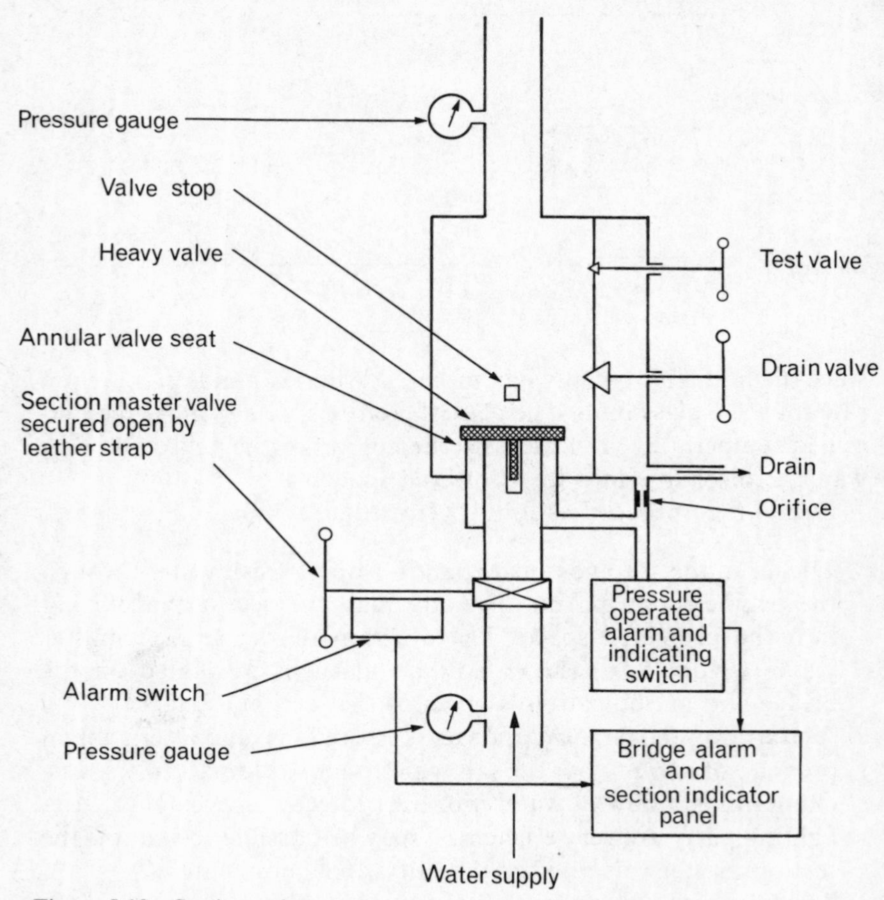

Figure 5.10 Section valve.

into the pressure-activated alarm switch, which instigates the bridge alarm and section indicator. The function of the orifice indicated is to allow small leaks past the heavy valve to pass to the drain, without their raising a false alarm. However, the orifice cannot cope with the sudden rush of water as the valve lifts, and the pressure immediately builds up to the value necessary to throw the pressure switch.

4. As water flows through each section valve, the engine room alarm should sound. This procedure is repeated through all the sections. The inspecting party usually includes personnel from both the deck and engineering departments.
5. The salt water pump (*see* Figure 5.9) is tested by closing the connection from the water main to the pressure sensor, and opening the small drain valve. The 'cut in' pressure is observed and the pump is allowed to discharge through the pump test valve.
6. After completion of the tests, the storage tank is replenished with water and air, and the system is returned to its normal fire-detection and fire-fighting role.

It will be noticed that the system will be effective, even in dry dock, because of the shore connection facility.

Chapter 6

Engineering Calculations

INDICATED POWER OF RECIPROCATING ENGINES

In Chapter 2, the ideal pressure–volume diagram for a diesel engine was considered (*see* Figure 2.18). In practice, such a diagram is not achieved; Figure 6.1 illustrates the actual diagram obtained in practice.

POWER DEVELOPED BY ONE PISTON

An *engine indicator mechanism* is used to obtain the diagram from the engine. It measures the variation in cylinder pressure with cylinder volume, as illustrated in Figure 6.2.

As explained in Chapter 2, the area of the diagram represents the energy transferred to the piston during one working cycle of the engine.

Energy transferred to piston during one working cycle = Constant which depends on the indicator mechanism used × Area of the actual indicator diagram taken from the engine

In order to make calculation easier, the area of the diagram is expressed in another way:

Area of diagram = Constant depending on the indicator mechanism × Mean height of diagram × Length of diagram

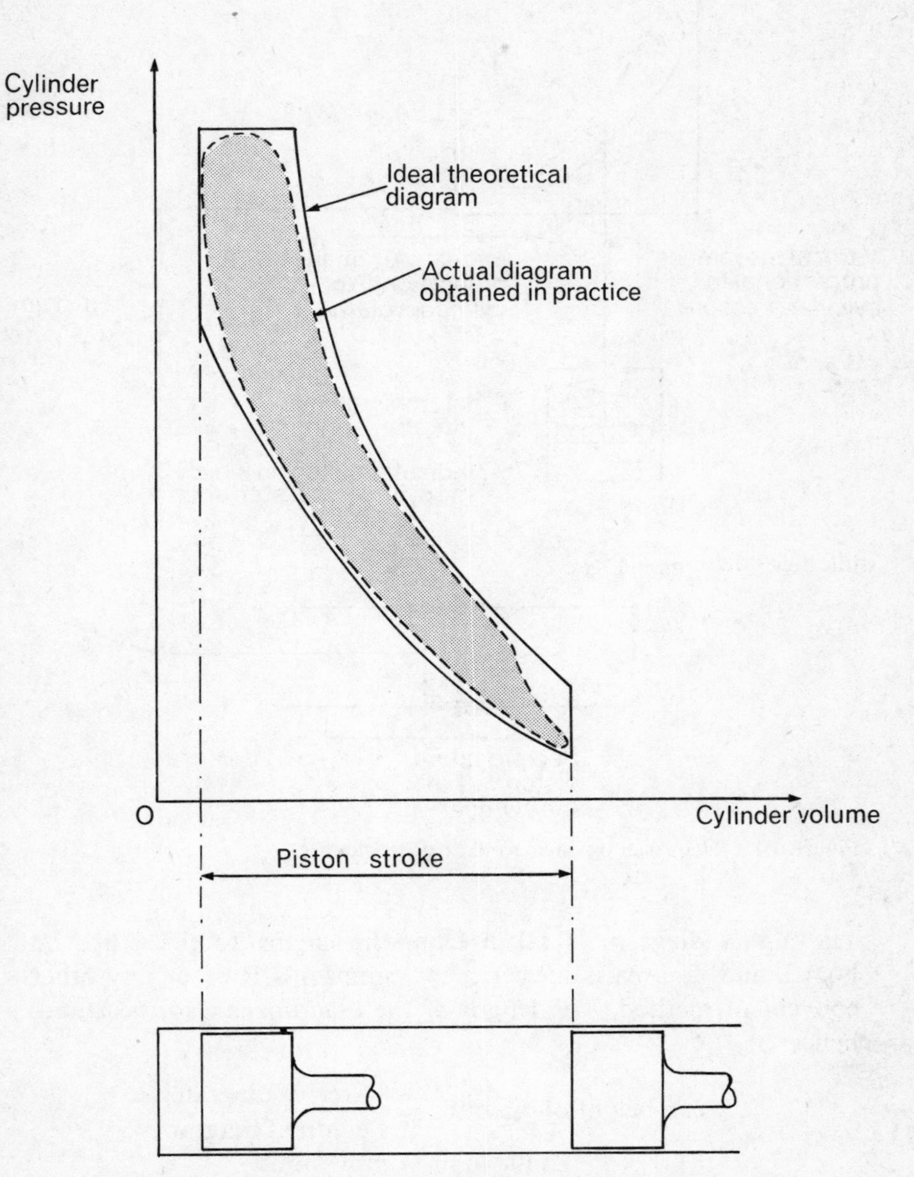

Figure 6.1 Ideal and actual indicator diagrams for a reciprocating engine.

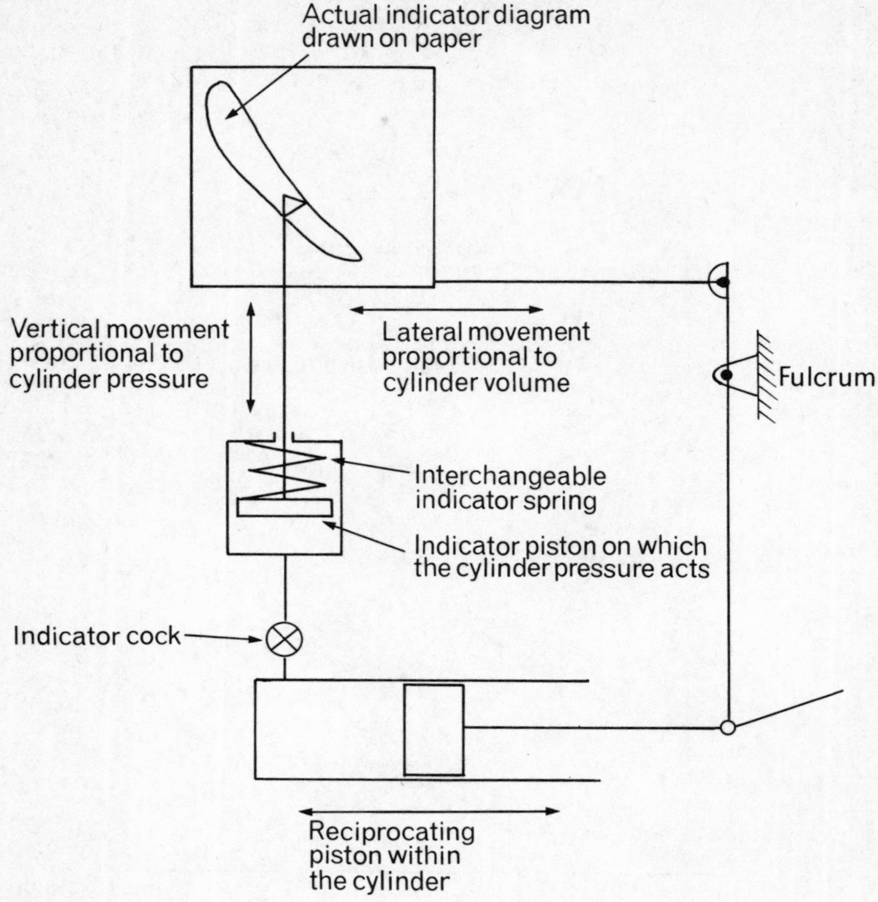

Figure 6.2 Reciprocating-engine indicator mechanism.

The actual diagram is taken from the engine to the office on board, and its area is measured by Simpson's Rule or any other convenient method. The length of the diagram is also measured; hence

$$\text{Mean height of diagram} = \frac{\text{Area of diagram}}{\text{Length of diagram}}$$

$$\text{Units: [mm} = \text{mm}^2/\text{mm]}$$

Since the height of the diagram is proportional to pressure, it follows that the mean height of the diagram is proportional to the mean effective cylinder pressure; hence

Mean effective cylinder pressure	=	Indicator constant	×	Mean height of actual diagram
Units: [N/m²	=	N/(m² mm)	×	mm]

Therefore

Energy transferred to piston during one working cycle	=	Mean effective cylinder pressure	×	Area of piston	×	Effective distance moved by piston
Units: [Nm	=	N/m²	×	m²	×	m]

Now, since power is defined as a rate of energy transfer,

Power developed by one piston	=	Mean effective cylinder pressure	×	Area of piston	×	Piston stroke	×	No. of working cycles of one cylinder per second
Units: [Nm/s	=	N/m²	×	m²	×	m	×	1/s]
	=	W (Watts)						

POWER DEVELOPED BY MULTICYLINDER ENGINE

From reference to the timing diagrams of the four-stroke and two-stroke engines in Figures 2.20 and 2.23, it can be seen that:

1. One cylinder of a four-stroke engine produces *one* complete working cycle for every *two* revolutions of the crankshaft; i.e.

$$\text{No. of working cycles per second} = \frac{\text{Revolutions per second}}{2}$$

2. One cylinder of a two-stroke engine produces *one* complete working cycle for every *one* revolution of the crankshaft; i.e.

$$\text{No. of working cycles per second} = \text{Revolutions per second}$$

To compute the power developed at the pistons of a multicylinder engine, the power of one cylinder is worked out, and then multiplied by the number of cylinders of the engine. An opposed piston engine is treated in the same way as any other engine, the combined piston stroke being used in the calculation. The power calculation obtained, following the procedure explained, is called the *indicated power*, because an *indicator mechanism* is used to obtain it. To summarize:

Indicated power of a multi-cylinder engine	=	Mean effective pressure acting on piston	×	Effective piston stroke length	×	Area of piston	×	Working cycles per second for one cylinder	×	No. of cylinders

Units:

$$\left[\begin{array}{l} \text{Nm/s} = \text{N/m}^2 \times \text{m}^2 \times \text{m} \times 1/\text{s} \\ \qquad\;\; = \text{W (Watts)} \end{array}\right]$$

Dividing throughout by 10^3 converts Watts to kilowatts. *Note:* The formula can be remembered as 'plan C'.

WORKED EXAMPLES

(1) An indicator diagram is taken from an engine and is divided into 10 equally spaced strips. The mid ordinates of the strips are 19, 21, 18, 15, 10, 8, 6, 4, 2·5, and 1·5 mm respectively. The spring constant is 60 $\text{kN/m}^2\text{mm}$. Find the mean effective pressure acting on the piston.

$$\text{Mean height of diagram} = \frac{\text{Total of ordinates}}{\text{No. of ordinates}}$$

$$= \frac{105}{10} = 10{\cdot}5 \text{ mm}$$

$$\text{Mean effective pressure} = \text{Mean height of diagram} \times \text{Indicator constant}$$

$$= 10{\cdot}5 \times 60 = 630 \text{ kN/m}^2$$

(2) The indicator diagram from a 6-cylinder 2-stroke engine has an area of 500 mm^2. The length of the diagram is 65 mm. The indicator constant is 63·7 $\text{kN/m}^2\text{mm}$. The swept volume of the engine is 0·25 m^3. Find:

(*a*) the mean effective pressure;

(*b*) the indicated power of the engine when it is revolving at 600 revolutions per minute.

Power of one cylinder:

$$\text{Mean effective pressure} = \underset{\text{of diagram}}{\text{Mean height}} \times \underset{\text{constant}}{\text{Indicator}}$$

$$= \frac{\text{Area of diagram}}{\text{Length of diagram}} \times \underset{\text{constant}}{\text{Indicator}}$$

$$= \frac{500}{65} \times 63{\cdot}7 = 490 \text{ kN/m}^2$$

$$\text{Units: } [\text{kN/m}^2 = \text{mm}^2/\text{mm} \times \text{kN/m}^2\text{mm}]$$

$$\underset{\text{power}}{\text{Indicated}} = \underset{\text{pressure}}{\overset{\text{Mean}}{\text{effective}}} \times \underset{\text{of stroke}}{\text{Length}} \times \underset{\text{piston}}{\text{Area of}} \times \underset{\text{second}}{\overset{\text{No. of working}}{\text{cycles per}}}$$

But

$$\text{Area of piston} \times \text{Length of stroke} = \text{Swept volume of piston}$$

Therefore

$$\text{Indicated power} = 490 \times 0{\cdot}25 \times 10 = 1\,225 \text{ kW}$$

$$\text{Units: } [\text{kW} = \text{kN/m}^2 \times \text{m}^3 \times 1/\text{s}]$$

Power of complete engine:

$$\text{Total indicated power} = \underset{\text{one cylinder}}{\text{Power output of}} \times \underset{\text{cylinders}}{\text{No. of}}$$

$$= 1\,225 \times 6$$

$$= 7\,350 \text{ kW}$$

BRAKE OR SHAFT POWER

The power developed in the engine cylinder of a reciprocating engine is transferred through mechanical devices to the propeller shaft. There are frictional losses due to these devices; hence the power measured at the propeller shaft is less than the indicated power.

Indicated power cannot be obtained directly from rotary machines, e.g. steam or gas turbines, and these plants rely on the measurement of shaft power. The shaft power is proportional to the torque transmitted along the shaft, which causes it to twist, as described earlier in Chapter 3.

Shaft power transmitted	=	Torque in shaft	×	Rotational speed of the shaft in radians per second
Units: [kW	=	kNm	×	l/s]

WORKED EXAMPLE

A propeller is absorbing a torque of 250 kNm and rotating at 60 revolutions per minute. Find the shaft power of the engine.

$$\text{Engine torque} = \text{Propeller shaft torque}$$

$$\text{Shaft power} = \text{Shaft torque} \times 2\pi \times \text{Revolutions per second}$$

$$= 250 \times 2\pi \times \frac{60}{60} = 1\,570{\cdot}8 \text{ kW}$$

$$\text{Units: } [\text{kNm/s} = \text{kNm} \times \text{l/s} = \text{kW}]$$

PROPELLER EFFICIENCY

The power transmitted through the propeller shaft, or the shaft power, is supplied to the propeller, which, due to its rotation,

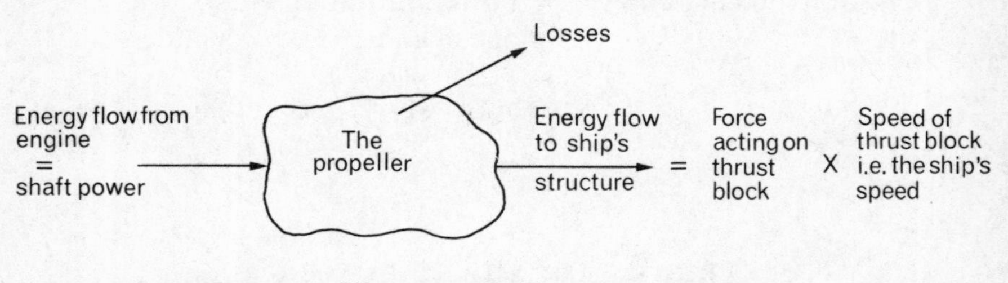

Figure 6.3 Propeller as a system for converting energy.

causes a thrust to act on the thrust block of the vessel, as illustrated in Figure 6.3. Hence

Effective power output of propeller	=	Force acting on thrust block	×	Speed of thrust block in the direction of the applied thrust force
Units: [kW	=	kN	×	m/s]

Hence the propeller efficiency can be defined thus:

$$\text{Propeller efficiency} = \frac{\text{Effective power output of the propeller}}{\text{Shaft power of the engine}}$$

WORKED EXAMPLE

A vessel is travelling at 16 knots in a straight line, at constant speed, when the shaft power of the engine is measured at 8 000 kW. Find the force transmitted to the thrust block, if the propeller efficiency is 70%.

$$\text{Propeller efficiency} = \frac{\text{Thrust force} \times \text{Ship speed}}{\text{Shaft power}}$$

1 knot is taken as 1 852 m/h = 1 852/3 600 m/s. Therefore

$$\text{Thrust force} = \frac{\text{Propeller efficiency} \times \text{Shaft power}}{\text{Ship speed}}$$

$$= \frac{70}{100} \times \frac{8\,000}{1} \times \frac{3\,600}{16 \times 1\,852} = 680{\cdot}35 \text{ kN}$$

$$\text{Units: [kN} = \quad \text{kNm/s} \times \text{s/m]}$$

(*Note:* 1 'old' British ton of force is very nearly equal to 10 kN.)

PROPELLER PITCH

The propeller generates a thrust force by changing the momentum of the stream of water passing through the propeller. The blades have an angle of incidence to the incoming water stream, which ensures that the optimum thrust force is obtained.

Consider a section of a propeller blade, at some radius from the shaft centre. The pitch of this blade section is defined as the distance this section would move forwards during one revolution, if it moved without slipping relative to the water. The pitch at a certain section can be defined thus, using Figure 6.4:

$$\text{Pitch at section} = 2\pi \times \text{Section radius} \times \text{Tangent of section pitch angle}$$

The pitches at different sections are probably different, and to find the pitch of the whole propeller, the pitch at each section is worked out and an average taken. Figure 6.5 illustrates a practical method of doing this.

WORKED EXAMPLE

The pitch angles of a propeller blade, measured at radii of 1·0, 2·0, 3·0, and 4·0, are respectively 45, 27, 20, and 16 degrees. Calculate the pitch of the propeller.

$$\text{Pitch at 1 m radius} = 2\pi \times 1 \times \tan 45^\circ = 6{\cdot}283\text{ m}$$
$$\text{Pitch at 2 m radius} = 2\pi \times 2 \times \tan 27^\circ = 6{\cdot}403\text{ m}$$
$$\text{Pitch at 3 m radius} = 2\pi \times 3 \times \tan 20^\circ = 6{\cdot}861\text{ m}$$
$$\text{Pitch at 4 m radius} = 2\pi \times 4 \times \tan 16^\circ = 7{\cdot}207\text{ m}$$
$$\text{Total} = 26{\cdot}754\text{ m}$$

Therefore

$$\text{Mean pitch of propeller} = \frac{\text{Total of sectional pitches}}{\text{No. of sections}}$$
$$= \frac{26{\cdot}754}{4} = 6{\cdot}6885\text{ m}$$

PROPELLER SLIP

The pitch of a propeller is the distance it would move forwards during one revolution if it rotated without slipping, like a nut on a bolt. However, propellers work in a fluid. Hence there will always be some slip relative to the water in which it is revolving, defined as:

$$\text{Propeller slip} = \frac{\begin{array}{c}\text{Theoretical forward}\\ \text{speed or distance}\\ \text{of the propeller}\end{array} - \begin{array}{c}\text{Actual forward}\\ \text{speed or distance}\\ \text{of the propeller}\end{array}}{\text{Theoretical forward speed or distance of the propeller}}$$

The *actual* forward speed of the propeller clearly is equal to the speed of the ship.

The *theoretical* forward speed can be worked out as follows:

$$\begin{array}{l}\text{Theoretical forward distance}\\ \text{moved during one revolution}\end{array} = \text{Pitch [m]}$$

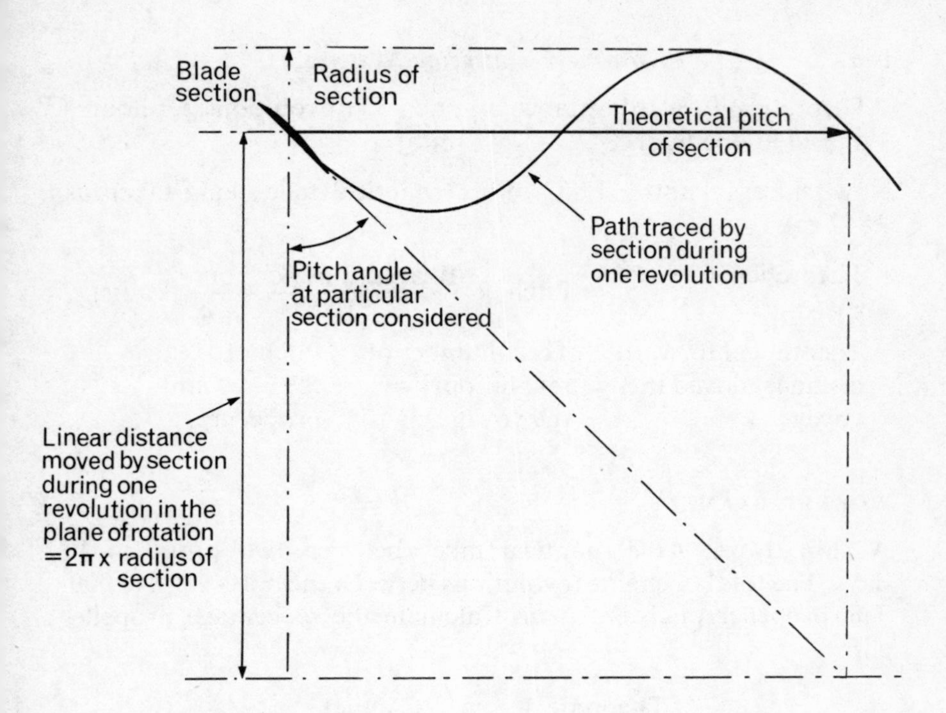

From the diagram,

$$\text{TAN.PITCH ANGLE OF SECTION} = \frac{\text{PITCH AT SECTION}}{2\pi \times \text{SECTION RADIUS}}$$

Figure 6.4 Theoretical definition of propeller pitch.

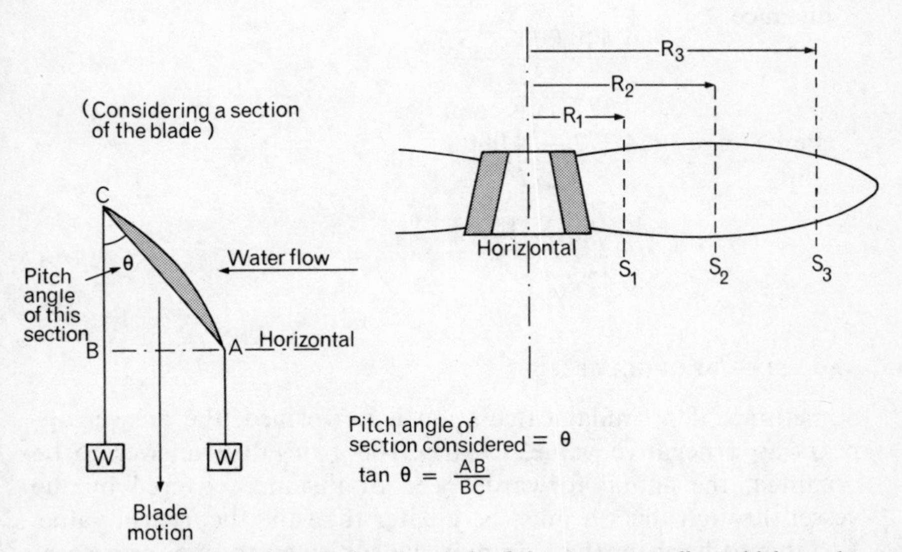

Figure 6.5 Method of measuring the pitch of the spare propeller, which can be used on board.

$$\text{Theoretical forward distance moved in one hour} = \text{Pitch} \times \text{Revolutions per hour} \quad [\text{m/h}]$$

Now, since 1 knot = 1 852 m/h (1 nautical mile being taken as 1 852 m),

$$\text{Theoretical forward speed} = \text{Pitch} \times \frac{\text{Revolutions per hour}}{1\,852} \quad [\text{knots}]$$

$$\text{Theoretical forward distance moved in a voyage} = \text{Total number of revolutions on the voyage} \times \text{Pitch of the propeller} \quad [\text{m}]$$

WORKED EXAMPLE

A ship travels 4 000 nautical miles between two ports, in 10 days. The total of engine revolutions between the ports is 1 440 000. The propeller pitch is 5·5 m. Calculate the percentage propeller slip.

$$\text{Percentage slip} = \frac{\text{Theoretical forward distance} - \text{Actual forward distance}}{\text{Theoretical forward distance}} \times 100\%$$

$$\begin{aligned}
\text{Theoretical forward distance} &= \frac{\text{Total revolutions} \times \text{Pitch}}{1\,852} \\
&= \frac{1\,440\,000 \times 5{\cdot}5}{1\,852} \\
&= 4\,277 \text{ nautical miles}
\end{aligned}$$

$$\begin{aligned}
\therefore \text{Percentage slip} &= \frac{4\,277 - 4\,000}{4\,277} \times 100\% \\
&= \frac{277}{4\,277} \times 100\% \\
&= 6{\cdot}477\%
\end{aligned}$$

NEGATIVE PROPELLER SLIP

Sometimes, if a similar calculation is performed, the answer appears as a negative value. Clearly, for a negative answer to be obtained, the actual forward speed or distance covered by the vessel through the sea must be greater than the theoretical value. For this to happen, the ship must be subjected to one, or a combination of, the following:

1. A following current.
2. A strong following wind.
3. A following sea.

LOG SLIP

The fundamental difference between a propeller and a propeller-driven log is that the propeller drives against the sea, but the sea drives the log. This means that a propeller-driven log must always have negative slip, if it is working in water which is moving at a relative velocity equal to the speed of the ship, i.e. in water which is not affected by the wake of the ship or by local speed fluctuations under the ship.

The log reading is the theoretical forward distance moved by the ship, which must be less than the actual distance run, provided that the stated conditions apply.

POWER REQUIRED TO PROPEL THE SHIP

Consider again Figure 6.3.

$$\text{Power developed by the propeller} = \text{Propeller thrust force} \times \text{Forward speed of the propeller} \quad (1)$$

If the ship is moving at a constant velocity, the propeller thrust force equals the sum of the following resistances:

1. Frictional resistances due to the flow of water past the wetted surface area of the ship.
2. Wave-making resistance, due to the loss of energy in lifting waves above the level of the sea.
3. Air resistance, due to the passage of air over the superstructure of the vessel.

The following introductory treatment to the power requirement of a ship, under different conditions, neglects wave-making and air resistance. Therefore,

$$\text{Propeller thrust force} = \text{Frictional resistances} \quad (2)$$

FRICTIONAL RESISTANCES

$$\text{Frictional resistances} \propto \text{Wetted surface area} \times (\text{Speed of ship})^2 \qquad (3)$$

On one particular ship, or in very similar 'sister' ships,

$$\text{Wetted surface area} \propto (\text{Physical dimensions of the ship})^2$$
$$\text{or } (WSA)^{\frac{1}{2}} \propto L \qquad (4)$$

It would be more convenient if *WSA* could be expressed in terms of the displacement of the vessel, where

$$\text{Displacement} \propto (\text{Physical dimensions of the ship})^3$$
$$\text{or } D^{\frac{1}{3}} \propto L$$

Then, from (4),

$$D^{\frac{1}{3}} \propto (WSA)^{\frac{1}{2}}$$
$$\text{or } D^{\frac{2}{3}} \propto WSA \qquad (5)$$

Therefore, from (3) and (5),

$$\text{Frictional resistances} \propto D^{\frac{2}{3}} \times \text{Speed}^2$$
$$\text{or } D^{\frac{2}{3}} \times V^2 \qquad (6)$$

From the introductory statements (1) and (2),

$$\text{Propeller power} \propto \text{Frictional resistance} \times \text{Speed}$$

Therefore, from (6),

$$\text{Propeller power} \propto D^{\frac{2}{3}} \times V^2 \times V$$
$$\text{or } D^{\frac{2}{3}} \times V^3 \qquad (7)$$

Within limits, it can be said that

$$\text{Propeller power} \propto \text{Engine power}$$

Hence, from (7),

$$\text{Engine power} \propto (\text{Displacement of ship})^{\frac{2}{3}} \times (\text{Speed of ship})^3$$
$$\text{or } P \propto D^{\frac{2}{3}} \times V^3 \qquad (8)$$

ADMIRALTY COEFFICIENT

Equating (8) to a constant,

$$\text{Constant, called the Admiralty coefficient for the particular ship considered} = \frac{D^{\frac{2}{3}} \times V^3}{P}$$

which is between 400 and 600 for average ships.

WORKED EXAMPLE

The power developed by a ship's engine was 10 000 kW, at a speed of 16 knots and a displacement of 40 000 tonnes. Estimate the power requirement to run at 12 knots and a displacement of 30 000 tonnes.

$$\text{Admiralty constant for this ship} = \frac{D_1^{\frac{2}{3}} V_1^{\,3}}{P_1} = \frac{D_2^{\frac{2}{3}} V_2^{\,3}}{P_2}$$

$$\frac{40\,000^{\frac{2}{3}} \times 16^3}{10\,000} = \frac{30\,000^{\frac{2}{3}} \times 12^3}{P_2}$$

$$P_2 = \left(\frac{30\,000}{40\,000}\right)^{\frac{2}{3}} \times \left(\frac{12}{16}\right)^3 \times 10\,000$$

$$= \frac{1}{1{\cdot}333^{\frac{2}{3}}} \times \frac{1}{1{\cdot}333^3} \times 10\,000$$

$$= \frac{1}{1{\cdot}211\,4} \times \frac{1}{2{\cdot}370} \times 10\,000$$

$$= 3\,483 \text{ kW}$$

FUEL CONSUMPTION

If it is assumed that the specific fuel consumption of an engine remains constant, for a range of engine powers, it can be said that

$$\text{Rate of fuel consumption} \propto \text{Power output of engine}$$
$$\text{or } RFC \propto P$$

Substituting in the formula for the Admiralty coefficient

$$\text{Constant, called the fuel coefficient} = \frac{D^{\frac{2}{3}} \times V^3}{RFC}$$

WORKED EXAMPLE

A vessel of 10 000-tonne displacement burns 20 tonnes of fuel per day at 14 knots. Estimate the fuel consumption for a 4 000-mile voyage at 16 knots, at a displacement of 8 000 tonnes.

$$\text{Fuel coefficient for this ship} = \frac{D_1^{\frac{2}{3}} V_1^3}{RFC_1} = \frac{D_2^{\frac{2}{3}} V_2^3}{RFC_2}$$

$$\frac{10\,000^{\frac{2}{3}} \times 14^3}{20} = \frac{8\,000^{\frac{2}{3}} \times 16^3}{RFC_2}$$

$$RFC_2 = \left(\frac{8\,000}{10\,000}\right)^{\frac{2}{3}} \times \left(\frac{16}{14}\right)^3 \times 20$$

$$= \frac{1}{1{\cdot}25^{\frac{2}{3}}} \times 1{\cdot}143^3 \times 20$$

$$= \frac{1}{1{\cdot}160\,4} \times 1{\cdot}493 \times 20$$

$$= 25{\cdot}727 \text{ tonnes per day}$$

$$\text{No. of days to complete the voyage} = \frac{4\,000}{24 \times 16}$$

$$= 10{\cdot}416 \text{ days}$$

$$\therefore \text{Total voyage consumption} = RFC_2 \times \text{No. of days}$$

$$= 25{\cdot}724 \times 10{\cdot}416$$

$$= 267{\cdot}927 \text{ tonnes}$$

Chapter 7

Reliability of Marine Systems

SYSTEMS FAILURE

To the ship handler, the essential requirement of the systems necessary for the safe control of the vessel is the ability of these systems to respond in a predictable way, with as high a reliability as can be attained under prevalent economic constraints.

It is suggested that, in order for the ship handler to gain confidence in the use of marine systems, he should be aware of the *limitations* inherent in these complex interdependent systems, especially those systems necessary for machinery spaces which are operated in the unattended condition. An important concept relating to the limitations of these systems, and therefore to the safety of these systems, is their *reliability*.

To engineers, the word 'reliability' has a special meaning. It has been defined as 'the probability that an item of equipment will perform a required function under specified conditions, without failure, for a specified period of time'. An engineer can never predict the precise elapsed time for the transition from a 'good' to a 'failed' state. However, he can gain information regarding the *probability* of such a transition after a certain elapsed time. Therefore reliability, and hence safety, are governed by the laws of probability, as well as by the knowledge and experience of the maintenance and operating personnel.

Consideration of the reliability of various systems leads to assertions regarding the probability of system failure in the very near future, described as the *failure rate*. Figure 7.1 illustrates various forms of failure rate.

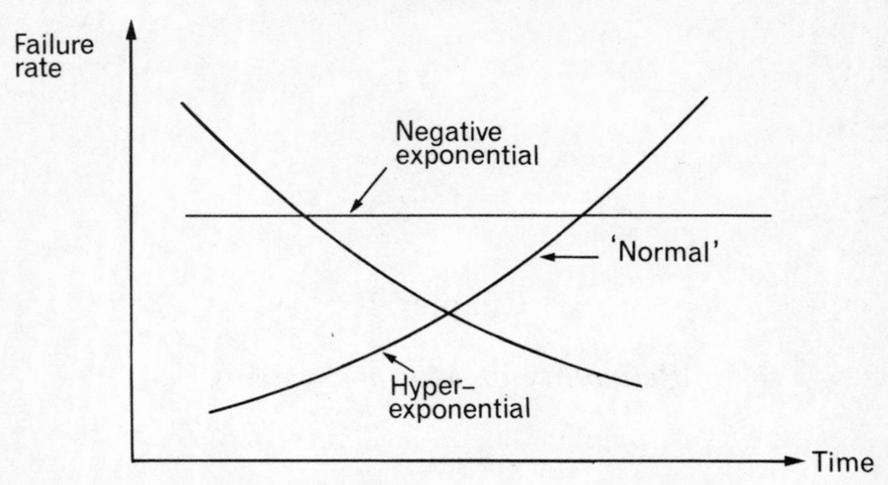

Figure 7.1 Various forms of failure rate.

1. The *normal* curve describes the failure rate of such things as electric light bulbs and small diesel engines.
2. The *negative exponential* curve describes generally the failure rate of systems with a great number of interdependent parts, where failure of any part causes transition from the good state of the complete system.
3. The *hyperexponential* curve applies to systems which have failure times which are either very long or very short. This curve has been found to describe the failure rate of electronic computers.

FAILURE RATE OF TOTAL SHIP SYSTEM

Marine installations broadly consist of the systems described in this book. Clearly these complex interdependent systems are composed of a very large number of single devices, each with its own failure rate. The envelope curve, illustrating the failure rate of the total ship system, is shown in Figure 7.2.

1. Stage 1 is called the *running-in, or burn-in, period*, and is characterized by the early failure of electronic components, by failures due to dirt in systems, and by mistakes in assembly.
2. Stage 2 illustrates the *normal operation* of the ship after it has been run in and tuned. The failure rate during this stage is at a

minimum but is not zero. Furthermore, it cannot be zero unless an infinite level of expenditure is contemplated—a point which is important to those engaged in ship operation. Failures will always occur in efficiently operated and maintained ships, due to chance failures in components which affect the response of the whole ship. The chance failures that do occur may be due to

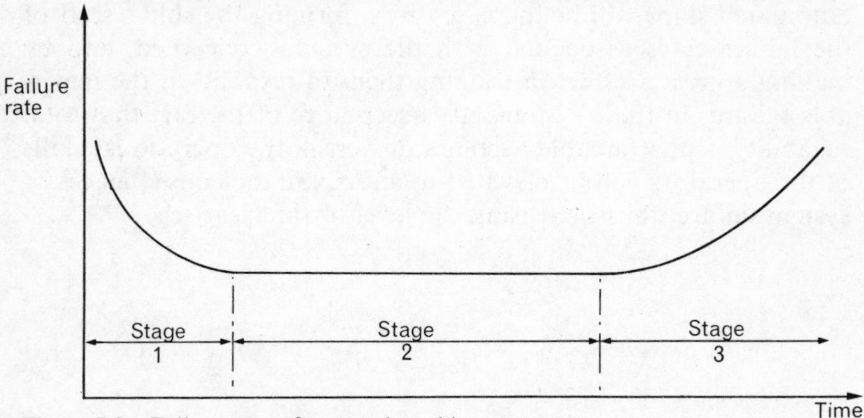

Figure 7.2 Failure rate of a complete ship system.

the low-probability chance combination of several adverse environmental factors, e.g. a rogue wave, or to a low-probability chance maldistribution of alloying elements in a component, resulting in its failure.

3. Stage 3 describes the *wear-out phase*, when a ship is approaching the end of its useful life.

CONCLUSIONS

Automated ships are designed and maintained by experts in their field and have been inspected and classified by a classification society.

The reliability of the systems involved has been improved by the knowledge gained, and the techniques developed, in the space programme of the US National Aeronautics and Space Administration. However, even in this programme, with the massive financial resources available, certain failures have occurred. For example, in the Saturn 5 and Apollo 8 which travelled to the moon in December 1968, there were only five failures of equipment, corresponding to the incredibly-high reliability level of

0·999999. The success of the NASA programme supports the idea of a high level of research and system development, combined with an education and training for the vehicle operators, which has been based on the expectancy of some equipment failure.

There must be common elements in the training of all varieties of vehicle operators, and it seems probable that the safety of automated ships will be increased by informing the ship's staff of the failure rates associated with the systems concerned, and by making a greater effort in training them to respond to the inevitable failures in these systems. As acceptance of the fact that total reliability is unattainable becomes universal, the operational skills of the operators will be elevated to anticipate the consequences of system failure, thus increasing the level of safety at sea.

Appendix

Specimen Examination Questions

MASTER (FOREIGN GOING)

ENGINEERING KNOWLEDGE, INSTRUMENTS AND CONTROL SYSTEMS

SPECIMEN PAPER 1 *Time allowed: 2 hours*

Candidates must attempt Question 1 and any *five* of the remainder.

1. Explain the sequence of events which occur in a vessel fitted with an automatic sprinkler system, from the time a fire starts in the accommodation area, to the time it is extinguished.
2. Sketch and describe a composite auxiliary boiler.
3. Describe the use of data loggers and mimic diagrams on board ship.
4. What is the function of a transducer? Describe a pressure–current transducer.
5. What is meant by 'supercharging' of a marine diesel engine, and why is this practice followed? Using a block diagram, show how the process is achieved.
6. Sketch and describe the operation of a resistance thermometer. Contrast the use of a thermistor in the system.
7. Explain the operation of an electromagnetic log, paying attention to the principles involved.
8. Discuss the relative merits of the analogue and digital methods of displaying and recording information.

MASTER (FOREIGN GOING)

ENGINEERING KNOWLEDGE, INSTRUMENTS AND CONTROL SYSTEMS

SPECIMEN PAPER 2 *Time allowed: 2 hours*

Candidates must attempt Question 1 and any *five* of the remainder.

1. A ship with a daily consumption of 35 tonnes at a speed of 15 knots left port with 350 tonnes of fuel on board. After 5 days' steaming at 15 knots, 550 miles are steamed at 16·5 knots to answer a distress call. Find the speed necessary to complete the remaining voyage distance of 1 200 miles, arriving with 35 tonnes in reserve. [Answer: 12·5 knots]
2. Using a block diagram of an automatic closed-loop control system, explain the operation of the system.
3. Sketch and describe a method of measuring the level in a tank, using an air purge system.
4. Define the terms 'indicated power' and 'brake power', explaining briefly how they are measured at sea.
5. Sketch and describe a regenerative condenser, explaining its particular operational advantages.
6. Explain, with the aid of sketches, the operating principle and installation of a thermocouple.
7. Using a sketch of the theoretical cycle diagram, explain the operation of a diesel engine.
8. Sketch an oil-fuel supply system for a boiler. Draw attention to any remotely operated components.

MASTER (FOREIGN GOING)

ENGINEERING KNOWLEDGE, INSTRUMENTS AND CONTROL SYSTEMS

SPECIMEN PAPER 3

Time allowed: 2 hours

Candidates must attempt Question 1 and any *five* of the remainder.

1. Estimate a dead-reckoning distance to use, given the following data:

Time	*Course*	*Log Reading*	*Engine Revolutions*	*Remarks*
1 200	075	0	109	Reduce revolutions
1 630	075	60	85	Reduce revolutions
1 730	075	72	50	Reduce revolutions
0 530	075	146	100	Increase revolutions
1 200	075	232	100	Increase revolutions

Log slip = −1·5% Propeller slip = +2% Propeller pitch = 4 m
[Answer: 233·657 nautical miles]

2. Draw a block diagram, illustrating the energy recovery systems used in a marine diesel engine.
3. Sketch a semi-rotary hydraulic proportional controller, paying particular attention to the feedback path.
4. Sketch and describe a flash evaporator, explaining its principle of operation.
5. Sketch and describe a rate of temperature rise detector, using bimetallic strips.
6. Using a block diagram, describe the operation of a bridge control system for a diesel engine.
7. Describe a float-operated tank-level measuring system. Use a sketch to illustrate your answer.
8. Draw a block diagram of a steering gear system. Explain how the output of the controller can be used to produce:

   (*a*) counter rudder control;
   (*b*) steering control;
   (*c*) automatic weather helm.

MASTER (FOREIGN GOING)

ENGINEERING KNOWLEDGE, INSTRUMENTS AND CONTROL SYSTEMS

SPECIMEN PAPER 4

Time allowed: 2 hours

Candidates must attempt Question 1 and any *five* of the remainder.

1. A vessel of 10 000 tonnes displacement consumes fuel at the rate of 75 tonnes per day at 18 knots, of which 7 tonnes per day are consumed by auxiliary plant. Calculate the bunkers required to complete a passage of 2 500 miles at 12 knots, allowing for 3 days' reserve for a speed of 18·75 knots. [Answer: 488·79 tonnes]
2. Explain the principle of operation of a radiation thermometer. Describe an instrument which can respond to radiations within a certain frequency band, giving an example of its use.
3. Describe how a steam turbine is manoeuvred manually. Give an example of the time from the 'full away' condition to the ship's coming to rest. Discuss the reasons for the time lags involved in the response curve.
4. Sketch and describe a remotely-operated cargo-valve system. Draw particular attention to the devices incorporated to ensure its safe operation.
5. Sketch a controllable pitch propeller, describing its method of operation.
6. Sketch and describe a transducer suitable for measuring stresses in a ship's hull. Using a block diagram, describe a hull surveillance system.
7. Draw a line diagram of a reciprocating pump, explaining its action. Explain the restrictions imposed on the suction lift of the pump.
8. Using a block diagram, explain the operation of a gas turbine power plant.

MASTER (FOREIGN GOING)

ENGINEERING KNOWLEDGE, INSTRUMENTS AND CONTROL SYSTEMS

SPECIMEN PAPER 5 *Time allowed: 2 hours*

Candidates must attempt Question 1 and any *five* of the remainder.

1. A four-bladed propeller has pitch angles of 40, 25, and 20 degrees, at radii of 0·5, 1·0, and 1·5 m respectively. It is fitted to a vessel which achieves a speed of 10·3 knots at 104 revolutions per minute. Calculate the slip of the propeller. [Answer: −1·98%] Explain negative slip, and the circumstances in which it occurs.
2. Explain the function and operation of the emergency generator, and how the machine would be started.
3. Sketch and describe a fire detection device which operates on the ionization principle.
4. Describe the operation of a centrifugal pump, with the aid of a sketch. Draw the discharge pressure–flow characteristic of the pump. Explain briefly how the suction conditions of the pump may be improved.
5. Describe the operation of a fire-detection and fire-fighting system using carbon dioxide. Differentiate clearly between the methods of fighting:

   (*a*) a fire in a cargo space;
   (*b*) a fire in an engine room.

6. Describe an electric telegraph based on the 'synchro' transmission system. List the alarms necessary for the safe operation of the system, explaining how manoeuvring of the engine is accomplished if the telegraph system is out of action.
7. Discuss, with the aid of a sketch, a system of remote hydraulic operation of hatch covers.
8. Explain briefly, with the aid of sketches:

   (*a*) a method of measuring pressure; and
   (*b*) a method of measuring flow.

MASTER (FOREIGN GOING)

ENGINEERING KNOWLEDGE, INSTRUMENTS AND CONTROL SYSTEMS

SPECIMEN PAPER 6 *Time allowed: 2 hours*

Candidates must attempt Question 1 and any *five* of the remainder.

1. With reference to a sketch of an inert-gas protection system for an oil tanker, describe briefly the operation of the system for:

   (*a*) loading cargo;
   (*b*) the loaded passage;
   (*c*) discharging;
   (*d*) ballasting;
   (*e*) tank cleaning.

2. Explain how proportional action and integral action are generated by a controller.
3. Sketch and describe a 'telemotor' transmission system. How would air, or leaks, in the system affect its action?
4. Describe the warming-through process for a steam turbine, after a protracted stay in port. Explain the reasons for following this procedure.
5. Using a block diagram, describe a bridge control system using a controllable pitch propeller.
6. Explain the principles of humidity measurement, and, using a sketch, describe the operation of an instrument to measure the humidity of the air in a cargo space.
7. A ship's engine is transmitting 10 000 kW to a propeller of efficiency 70%. If the vessel is proceeding at 15 knots, calculate the total resistive forces acting on the vessel, discussing their origin. [Answer: 907·127 kN]
8. Draw a block diagram of a refrigeration system, explaining its operation. Briefly draw attention to the properties required of:

   (*a*) primary refrigerants;
   (*b*) secondary refrigerants.

MASTER (FOREIGN GOING)

ENGINEERING KNOWLEDGE, INSTRUMENTS AND CONTROL SYSTEMS

SPECIMEN PAPER 7

Time allowed: 2 hours

Candidates must attempt Question 1 and any *five* of the remainder.

1. A vessel loads bunkers for a voyage of 4 600 miles at 15 knots. The main engine consumes 49 tonnes per day at this speed, the auxiliary plant requiring 3 tonnes per day, which is independent of speed. Just before sailing at 1824 hours on 3 March, the vessel is ordered to proceed on the voyage at 13·5 knots. At 0624 hours on 10 March, orders are received to proceed for the rest of the voyage at 15·4 knots. Calculate:

   (*a*) the estimated time of arrival, [Answer: 0021 hours on 17 March]
   (*b*) the bunkers remaining at arrival. [Answer: 30·273 tonnes]
2. Using a block diagram, illustrate the flows of information relating to a machinery control room. Explain the requirements at the bridge position for a vessel operating under U.M.S. conditions, and describe the role of the deck officer in this mode of operation.
3. Draw a graph illustrating the general flammability limits of a hydrocarbon gas and air mixture. Indicate significant areas on the graph, with approximate percentage values. Sketch and describe an instrument for detecting the presence of hydrocarbons.
4. Describe a system which allows a main cargo pump to be operated as a stripping pump.
5. Describe, with the aid of a block diagram, the operation of a fin-operated stabilization system.
6. Draw a sketch of a feed water system for a steam plant. Very briefly, state the function of each component in the feed system, as you describe the system.
7. Describe the operation of a two-stroke diesel engine, with particular reference to the scavenging process.
8. Describe, with the aid of sketches, the operating principle and the arrangement of an oil-in-water detection system.

MASTER (FOREIGN GOING)

ENGINEERING KNOWLEDGE, INSTRUMENTS AND CONTROL SYSTEMS

SPECIMEN PAPER 8

Time allowed: 2 hours

Candidates must attempt Question 1 and any *five* of the remainder.

1. A ship of 5 000 tonnes displacement, with a fuel consumption rate of 25 tonnes per day at 15 knots, leaves with 500 tonnes of fuel on board, for a port 3 600 miles distant. After discharging 2 500 tonnes of cargo in 5 days, orders are received to proceed to a loading port, 1 200 miles away at 10 knots. How much fuel is left on arrival at this port, assuming the fuel consumption rate for auxiliary plant is constant at 4 tonnes per day? [Answer: 148·64 tonnes]
2. Explain how a standby generator may be automatically started and connected to the electrical distribution system.
3. Draw a sketch of a main condenser and its connections. Describe the circumstances in which the various valves in the system may be used, paying particular attention to any emergency use.
4. Describe an inert-gas generating and distribution system, which could be used for the detection and extinguishing of a fire in a cargo hold. Draw a graph showing the reduction of oxygen in a cargo space, on a time base.
5. Draw a line diagram of a water-tube boiler, describing its operation. Briefly explain the reasons for the necessity for a warming-up period, when steam is being raised from a cold boiler.
6. Briefly describe, with simple sketches, the operation of:

   (*a*) an axial flow pump;
   (*b*) a positive-displacement rotary pump.

   Illustrate the discharge pressure–flow characteristic of each type of pump.
7. Sketch and describe a pneumatic transmitter. Could this device be used as a controller, and, if so, what kind of control action would it generate?
8. Describe a device which could be used to measure a ship's draught, illustrating your answer with a sketch.

MASTER (FOREIGN GOING)

ENGINEERING KNOWLEDGE, INSTRUMENTS AND CONTROL SYSTEMS

SPECIMEN PAPER 9 *Time allowed: 2 hours*

Candidates must attempt Question 1 and any *five* of the remainder.

1. A vessel displacing 109 500 tonnes has 2 790 tonnes of fuel to consume, with a reserve of 500 tonnes. Its rate of fuel consumption is 93 tonnes per day at 16·5 knots. On a voyage of 11 520 miles, and after 12 days' steaming at 12 knots, 700 tonnes of fuel are found to be so contaminated that they cannot be used. Find the speed at which the voyage should be completed in order for the ship to have 500 tonnes in reserve, on arrival. [Answer: 15·45 knots]
2. Using a block diagram, describe the operation of the bridge control system for a steam turbine. Explain briefly the effect of the power-plant protective devices on the response of the plant.
3. Draw a block diagram of an automatic closed-loop steering system. List those adjustments that can be made to the controller, and briefly describe their effect on the response of the vessel.
4. Draw a line diagram of a 'Scotch' marine boiler. Explain, giving reasons, the necessity of allowing a long period for raising steam from cold, in such a boiler.
5. Describe a power plant installation using medium-speed diesel engines, with a controllable pitch propeller. Briefly describe other methods of obtaining astern thrust.
6. Sketch and briefly describe:

   (*a*) a sprinkler head for a fire-detection and fire-fighting system;
   (*b*) a bilge level alarm.

7. Many ships are fitted with a comprehensive data-acquisition and alarm system. Illustrate such a system, using a block diagram. Explain the function and operation of the system.
8. Explain the necessity for producing distilled water for making up the water losses from a feed system. Sketch and briefly describe a type of boiling evaporator which provides such water.

Index